TensorFlow 2
Pocket Reference
Building and Deploying
Machine Learning Models

KC Tung

Beijing · Boston · Farnham · Sebastopol · Tokyo

TensorFlow 2 Pocket Reference

by KC Tung

Published by O'Reilly Media, Inc., 1005 Gravenstein Highway North, Sebastopol, CA 95472.

O'Reilly books may be purchased for educational, business, or sales promotional use. Online editions are also available for most titles (*http://oreilly.com*). For more information, contact our corporate/institutional sales department: 800-998-9938 or *corporate@oreilly.com*.

Acquisitions Editor: Rebecca Novack
Development Editor: Sarah Grey
Production Editor: Beth Kelly
Copyeditor: Penelope Perkins
Proofreader: Audrey Doyle
Indexer: Potomac Indexing, LLC
Interior Designer: David Futato
Cover Designer: Karen Montgomery
Illustrator: Kate Dullea

August 2021:   First Edition

Revision History for the First Edition
        2021-07-19:   First Release

See *https://oreil.ly/tf2pr* for release details.

978-1-492-08918-6

[LSI]

To my beloved wife Katy, who always supports me and sees the best in me. To my father Jerry, who raised me to pursue learning with a sense of purpose. To my hard-working and passionate readers, whose aspiration for continuous learning resonates with me and inspired me to write this book.

Table of Contents

Preface

The TensorFlow ecosystem has evolved into many different frameworks to serve a variety of roles and functions. That flexibility is part of the reason for its widespread adoption, but it also complicates the learning curve for data scientists, machine learning (ML) engineers, and other technical stakeholders. There are so many ways to manage TensorFlow models for common tasks—such as data and feature engineering, data ingestions, model selection, training patterns, cross validation against overfitting, and deployment strategies—that the choices can be overwhelming.

This pocket reference will help you make choices about how to do your work with TensorFlow, including how to set up common data science and ML workflows using TensorFlow 2.0 design patterns in Python. Examples describe and demonstrate TensorFlow coding patterns and other tasks you are likely to encounter frequently in the course of your ML project work. You can use it as both a how-to book and a reference.

This book is intended for current and potential ML engineers, data scientists, and enterprise ML solution architects who want to advance their knowledge and experience in reusable patterns and best practices in TensorFlow modeling. Perhaps you've already read an introductory TensorFlow book, and you stay up to date with the field of data science generally. This book assumes that you have hands-on experience using Python (and possibly NumPy, pandas, and JSON libraries) for data engineering, feature engineering routines, and building TensorFlow models. Experience with common data structures such as lists, dictionaries, and NumPy arrays will also be very helpful.

Unlike many other TensorFlow books, this book is structured around the tasks you'll likely need to do, such as:

- When and why should you feed training data as a NumPy array or streaming dataset? (Chapters 2 and 5)
- How can you leverage a pretrained model using transfer learning? (Chapters 3 and 4)
- Should you use a generic fit function to do your training or write a custom training loop? (Chapter 6)
- How should you manage and make use of model checkpoints? (Chapter 7)
- How can you review the training process using Tensor-Board? (Chapter 7)
- If you can't fit all of your data into your runtime's memory, how can you perform distributed training using multiple accelerators, such as GPUs? (Chapter 8)
- How do you pass data to your model during inferencing and how do you handle output? (Chapter 9)
- Is your model fair? (Chapter 10)

If you are wrestling with questions like these, this book will be helpful to you.

Conventions Used in This Book

The following typographical conventions are used in this book:

Italic
Indicates new terms, URLs, email addresses, filenames, and file extensions.

Constant width

> Used for program listings, as well as within paragraphs to refer to program elements such as variable or function names, databases, data types, environment variables, statements, and keywords.

Constant width bold

> Shows commands or other text that should be typed literally by the user.

Constant width italic

> Shows text that should be replaced with user-supplied values or by values determined by context.

TIP

This element signifies a tip or suggestion.

Using Code Examples

Supplemental material (code examples, exercises, etc.) can be downloaded at *https://github.com/shinchan75034/tensorflow-pocket-ref*.

If you have a technical question or a problem using the code examples, please send email to *bookquestions@oreilly.com*.

This book is here to help you get your job done. In general, if example code is offered with this book, you may use it in your programs and documentation. You do not need to contact us for permission unless you're reproducing a significant portion of the code. For example, writing a program that uses several chunks of code from this book does not require permission. Selling or distributing examples from O'Reilly books does require permission. Answering a question by citing this book and quoting example code does not require permission. Incorporating a significant amount of example code from this

---

book into your product's documentation does require permission.

We appreciate, but generally do not require, attribution. An attribution usually includes the title, author, publisher, and ISBN. For example: "*TensorFlow 2 Pocket Reference* by KC Jung (O'Reilly). Copyright 2021 Favola Vera, LLC, 978-1-492-08918-6."

If you feel your use of code examples falls outside of fair use or the permission given above, feel free to contact us at *permissions@oreilly.com*.

O'Reilly Online Learning

 For more than 40 years, *O'Reilly Media* has provided technology and business training, knowledge, and insight to help companies succeed.

Our unique network of experts and innovators share their knowledge and expertise through books, articles, and our online learning platform. O'Reilly's online learning platform gives you on-demand access to live training courses, in-depth learning paths, interactive coding environments, and a vast collection of text and video from O'Reilly and 200+ other publishers. For more information, visit *http://oreilly.com*.

How to Contact Us

Please address comments and questions concerning this book to the publisher:

O'Reilly Media, Inc.
1005 Gravenstein Highway North
Sebastopol, CA 95472
800-998-9938 (in the United States or Canada)
707-829-0515 (international or local)
707-829-0104 (fax)

We have a web page for this book, where we list errata, examples, and any additional information. You can access this page at *https://oreil.ly/tensorflow2pr*.

Email *bookquestions@oreilly.com* to comment or ask technical questions about this book.

For news and information about our books and courses, visit *http://oreilly.com*.

Find us on Facebook: *http://facebook.com/oreilly*

Follow us on Twitter: *http://twitter.com/oreillymedia*

Watch us on YouTube: *http://youtube.com/oreillymedia*

Acknowledgments

I really appreciate all the thoughtful and professional works by O'Reilly editors. In addition, I also want to express my gratitude to technical reviewers: Tony Holdroyd, Pablo Marin, Giorgio Saez, and Axel Sirota for their valuable feedback and suggestions. Finally, a special thank to Rebecca Novack and Sarah Grey for giving me a chance and working with me to write this book.

## CHAPTER 1

Introduction to TensorFlow 2

TensorFlow has long been the most popular open source Python machine learning (ML) library. It was developed by the Google Brain team as an internal tool, but in 2015 it was released under an Apache License. Since then, it has evolved into an ecosystem full of important assets for model development and deployment. Today it supports a wide variety of APIs and modules that are specifically designed to handle tasks such as data ingestion and transformation, feature engineering, and model construction and serving, as well as many more.

TensorFlow has become increasingly complex. The purpose of this book is to help simplify the common tasks that a data scientist or ML engineer will need to perform during an end-to-end model development process. This book does not focus on data science and algorithms; rather, the examples here use pre-built models as a vehicle to teach relevant concepts.

This book is written for readers with basic experience in and knowledge about building ML models. Some proficiency in Python programming is highly recommended. If you work through the book from beginning to end, you will gain a great deal of knowledge about the end-to-end model development process and the major tasks involved, including data engineering, ingestion, and preparation; model training; and serving the model.

The source code for the examples in the book was developed and tested with Google Colaboratory (Colab, for short) and a MacBook Pro running macOS Big Sur, version 11.2.3. The TensorFlow version used is 2.4.1, and the Python version is 3.7.

Improvements in TensorFlow 2

As TensorFlow grows, so does its complexity. The learning curve for new TensorFlow users is steep because there are so many different aspects to keep in mind. *How do I prepare the data for ingestion and training? How do I handle different data types? What do I need to consider for different handling methods?* These are just some of the basic questions you may have early in your ML journey.

A particularly difficult concept to get accustomed to is *lazy execution*, which means that TensorFlow doesn't actually process your data until you explicitly tell it to execute the entire code. The idea is to speed up performance. You can look at an ML model as a set of nodes and edges (in other words, a graph). When you run computations and transform data through the nodes in the path, it turns out that only the computations in the datapath are executed. In other words, you don't have to calculate every computation, only the ones that lie directly in the path your data takes through the graph from input through output. If the shape and format of the data are not correctly matched between one node and the next, when you compile the model you will get an error. It is rather difficult to investigate where you made a mistake in passing a data structure or tensor shape from one node to the next to debug.

Through TensorFlow 1.*x*, lazy execution was the way to build and train an ML model. Starting with TensorFlow 2, however, *eager execution* is the default way to build and train a model. This change makes it much easier to debug the code and try different model architectures. Eager execution also makes it much easier to learn TensorFlow, in that you will see any mistakes immediately upon executing each line of code. You no longer need to build an entire graph of your model before you can debug and test whether your input data is in the right shape. This is one of several major features and improvements that make TensorFlow 2 easier to use than previous versions.

Keras API

Keras, created by AI researcher François Chollet, is an open source, high-level, deep-learning API or framework. It is compatible with multiple ML libraries.

High-level implies that at a lower level there is another framework that actually executes the computation—and this is indeed the case. These low-level frameworks include TensorFlow, Theano, and the Microsoft Cognitive Toolkit (CNTK). The purpose of Keras is to provide easier syntax and coding style for users who want to leverage the low-level frameworks to build deep-learning models.

After Chollet joined Google in 2015, Keras gradually became a keystone of TensorFlow adoption. In 2019, as the TensorFlow team launched version 2.0, it formally adopted Keras as TensorFlow's first-class citizen API, known as `tf.keras`, for all future releases. Since then, TensorFlow has integrated `tf.keras` with many other important modules. For example, it works seamlessly with the `tf.io` API for reading distributed training data. It also works with the `tf.data.Dataset` class, used for streaming training data too big to fit into a single computer. This book uses these modules throughout all chapters.

Today TensorFlow users primarily rely on the `tf.keras` API for building deep models quickly and easily. The convenience of getting the training routine working quickly allows more time to experiment with different model architectures and tuning parameters in the model and training routine.

Reusable Models in TensorFlow

Academic researchers have built and tested many ML models, all of which tend to be complicated in their architecture. It is not practical for users to learn how to build these models. Enter the idea of *transfer learning*, where a model developed for one task is reused to solve another task, in this case one defined by the user. This essentially boils down to transforming user data into the proper data structure at model input and output.

Naturally, there has been great interest in these models and their potential uses. Therefore, by popular demand, many models have become available in the open source ecosystem. TensorFlow created a repository, TensorFlow Hub, to offer the public free access to these complicated models. If you're interested, you can try these models without having to build them yourself. In Chapter 4, you will learn how to download and use models from TensorFlow Hub. Once you do, you'll just need to be aware of the data structure the model expects at input, and add a final output layer that is suitable for your prediction goal. Every model in TensorFlow Hub contains concise documentation that gives you the necessary information to construct your input data.

Another place to retrieve prebuilt models is the tf.keras.applications module, which is part of the TensorFlow distribution. In Chapter 4, you'll learn how to use this module to leverage a prebuilt model for your own data.

Making Commonly Used Operations Easy

All of these improvements in TensorFlow 2 make a lot of important operations easier and more convenient to implement. Even so, building and training an ML model end to end is not a trivial task. This book will show you how to deal with each aspect of the TensorFlow 2 model training process, starting from the beginning. Following are some of these operations.

Open Source Data

A convenient package integrated into TensorFlow 2 is the
TensorFlow dataset library (*https://oreil.ly/0nt9T*). It is a collection of curated open source datasets that are readily available
for use. This library contains datasets of images, text, audio,
videos, and many other formats. Some are NumPy arrays, while
others are in dataset structures. This library also provides documentation for how to use TensorFlow to load these datasets.
By distributing a wide variety of open source data with its
product, the TensorFlow team really saves users a lot of the
trouble of searching for, integrating, and reshaping training
data for a TensorFlow workload. Some of the open source datasets we'll use in this book are the *Titanic* dataset (*https://oreil.ly/
GWCN1*) for structured data classification and the CIFAR-10
dataset (*https://oreil.ly/uwQUm*) for image classification.

Working with Distributed Datasets

First you have to deal with the question of how to work with
training data. Many didactic examples teach TensorFlow using
prebuilt training data in its native format, such as a small pandas DataFrame or a NumPy array, which will fit nicely in your
computer's memory. In a more realistic situation, however,
you'll likely have to deal with much more training data than
your computer memory can handle. The size of a table read
from a SQL database can easily reach into the gigabytes. Even if
you have enough memory to load it into a pandas DataFrame
or a NumPy array, chances are your Python runtime will run
out of memory during computation and crash.

Large tables of data are typically saved as multiple files in common formats such as CSV (comma-separated value) or text.
Because of this, you should not attempt to load each file in your
Python runtime. The correct way to deal with distributed datasets is to create a reference that points to the location of *all* the
files. Chapter 2 will show you how to use the tf.io API, which
gives you an object that holds a list of file paths and names.

This is the preferred way to deal with training data regardless of its size and file count.

Data Streaming

How do you intend to pass data to your model for training? This is an important skill, but many popular didactic examples approach it by passing the entire NumPy array into the model training routine. Just like with loading large training data, you will encounter memory issues if you try passing a large NumPy array to your model for training.

A better way to deal with this is through *data streaming*. Instead of passing the entire training data at once, you stream a subset or batch of data for the model to train with. In Tensor-Flow, this is known as your *dataset*. In Chapter 2, you are also going to learn how to make a dataset from the `tf.io` object. Dataset objects can be made from all sorts of native data structures. In Chapter 3, you will see how to make a `tf.data.Data set` object from CSV files and images.

With the combination of `tf.io` and `tf.data.Dataset`, you'll set up a data handling workflow for model training without having to read or open a single data file in your Python runtime memory.

Data Engineering

To make meaningful features for your model to learn the pattern of, you need to apply data- or feature-engineering tasks to your training data. Depending on the data type, there are different ways to do this.

If you are working with tabular data, you may have different values or data types in different columns. In Chapter 3, you will see how to use TensorFlow's `feature_column` API to standardize your training data. It helps you correctly mark which columns are numeric and which are categorical.

For image data, you will have different tasks. For example, all of the images in your dataset must have the same dimensions. Further, pixel values are typically normalized or scaled to a range of [0, 1]. For these tasks, tf.keras provides the ImageDataGenerator class, which standardizes image sizes and normalizes pixel values for you.

Transfer Learning

TensorFlow Hub makes prebuilt, open source models available to everyone. In Chapter 4, you'll learn how to use the Keras layers API to access TensorFlow Hub. In addition, tf.keras comes with an inventory of these prebuilt models, which can be called using the tf.keras.applications module. In Chapter 4, you'll learn how to use this module for transfer learning as well.

Model Styles

There is definitely more than one way you can implement a model using tf.keras. This is because some deep learning model architectures or patterns are more complicated than others. For common use, the *symbolic API* style, which sets up your model architecture sequentially, is likely to suffice. Another style is *imperative API*, where you declare a model as a class, so that each time you call upon a model object, you are creating an instance of that class. This requires you to understand how class inheritance works (I'll discuss this in Chapter 6). If your programming background stems from an object-oriented programming language such as C++ or Java, then this API may have a more natural feel for you. Another reason for using the imperative API approach is to keep your model architecture code separate from the remaining workflow. In Chapter 6, you will learn how to set up and use both of these API styles.

Monitoring the Training Process

Monitoring how your model is trained and validated across each *epoch* (that is, one pass over a training set) is an important aspect of model training. Having a validation step at the end of each epoch is the easiest thing you can do to guard against *model overfitting*, a phenomenon in which the model starts to memorize training data patterns rather than learning the features as intended. In Chapter 7, you will learn how to use various *callbacks* to save model weights and biases at every epoch. I'll also walk you through how to set up and use Tensor-Board to visualize the training process.

Distributed Training

Even though you know how to handle distributed data and files and stream them into your model training routine, what if you find that training takes an unrealistic amount of time? This is where *distributed training* can help. It requires a cluster of hardware accelerators, such as graphics processing units (GPUs) or Tensor Processing Units (TPUs). These accelerators are available through many public cloud providers. You can also work with one GPU or TPU (not a cluster) for free in Google Colab; you'll learn how to use this and the `tf.distribute.MirroredStrategy` class, which simplifies and reduces the hard work of setting up distributed training, to work through the example in the first part of Chapter 8.

Released before `tf.distribute.MirroredStrategy`, the Horovod API from Uber's engineering team is a considerably more complicated alternative. It's specifically built to run training routines on a computing cluster. To learn how to use Horovod, you will need to use Databricks, a cloud-based computing platform, to work through the example in the second part of Chapter 8. This will help you learn how to refactor your code to distribute and shard data for the Horovod API.

Serving Your TensorFlow Model

Once you've built your model and trained it successfully, it's time for you to persist, or store, the model so it can be served to handle user input. You'll see how easy it is to use the `tf.saved_model` API to save your model.

Typically, the model is hosted by a web service. This is where TensorFlow Serving comes into the picture: it's a framework that wraps your model and exposes it for web service calls via HTTP. In Chapter 9, you will learn how to use a TensorFlow Serving Docker image to host your model.

Improving the Training Experience

Finally, Chapter 10 discusses some important aspects of assessing and improving your model training process. You'll learn how to use the TensorFlow Model Analysis module to look into the issue of model bias. This module provides an interactive dashboard, called Fairness Indicators, designed to reveal model bias. Using a Jupyter Notebook environment and the model you trained on the *Titanic* dataset from Chapter 3, you'll see how Fairness Indicators works.

Another improvement brought about by the `tf.keras` API is that it makes performing hyperparameter tuning more convenient. *Hyperparameters* are attributes related to model training routines or model architectures. Tuning them is typically a tedious process, as it involves thoroughly searching over the parameter space. In Chapter 10 you'll see how to use the Keras Tuner library and an advanced search algorithm known as Hyperband to conduct hyperparameter tuning work.

Wrapping Up

TensorFlow 2 is a major overhaul of the previous version. Its most significant improvement is designating the `tf.keras` API as the recommended way to use TensorFlow. This API works seamlessly with `tf.io` and `tf.data.Dataset` for an end-to-end model training process. These improvements speed up model building and debugging so you can experiment with other aspects of model training, such as trying different architectures or conducting more efficient hyperparameter searches. So, let's get started.

## CHAPTER 2

Data Storage and Ingestion

To envision how to set up an ML model to solve a problem, you have to start thinking about data structure patterns. In this chapter, we'll look at some general patterns in storage, data formats, and data ingestion. Typically, once you understand your business problem and set it up as a data science problem, you have to think about how to get the data into a format or structure that your model training process can use. Data ingestion during the training process is fundamentally a data transformation pipeline. Without this transformation, you won't be able to deliver and serve the model in an enterprise-driven or use-case-driven setting; it would remain nothing more than an exploration tool and would not be able to scale to handle large amounts of data.

This chapter will show you how to design a data ingestion pipeline for two common data structures: tables and images. You will learn how to make the pipeline scalable by using TensorFlow's APIs.

Data streaming is the means by which the data is ingested in small batches by the model for training. Data streaming in Python is not a new concept. However, grasping it is fundamental to understanding how the more advanced APIs in TensorFlow work. Thus, this chapter will start with Python generators. Then we'll look at how tabular data is stored, including how to indicate and track features and labels. We'll then move to designing your data structure, and finish by discussing how to ingest data to your model for training and how to stream tabular data. The rest of the chapter covers how to organize image data for image classification and how to stream image data.

Streaming Data with Python Generators

There are times when the Python runtime's memory is not big enough to handle loading the dataset in its entirety. When this happens, the recommended practice is to load the data in small batches. Therefore, the data is streamed into the model during the training process.

Sending data in small batches has many other advantages as well. One is that a gradient descent algorithm is applied to each batch to calculate the error (that is, the difference between the model output and the ground truth) and to gradually update the model's weights and biases to make this error as small as possible. This lets us parallelize the gradient calculation, since the error calculation (also known as *loss calculation*) of one batch does not depend on the other. This is known as *mini-batch gradient descent*. At the end of each epoch, after a full training dataset has gone through the model, gradients from all batches are summed and weights are updated. Then, training starts again for the next epoch, with the newly updated weights and biases, and the error is calculated. This process repeats according to a user-defined parameter, which is known as *number of epochs for training*.

A *Python generator* is an iterator that returns an iterable. An example of how it works follows. Let's start with a NumPy library for this simple demonstration of Python generators. I've created a function, my_generator, that accepts a NumPy array and iterates two records at a time in the array:

```
import numpy as np

def my_generator(my_array):
    i = 0
    while True:
        yield my_array[i:i+2, :] # output two elements at a time
        i += 1
```

This is the test array I created, which will be passed into
my_generator:

```
test_array = np.array([[10.0, 2.0],
                       [15, 6.0],
                       [3.2, -1.5],
                       [-3, -2]], np.float32)
```

This NumPy array has four records, each consisting of two
floating-point values. Then I pass this array to my_generator:

```
output = my_generator(test_array)
```

To get output, use:

```
next(output)
```

The output should be:

```
array([[10.,  2.],
       [15.,  6.]], dtype=float32)
```

If you run the next(output) command again, the output will be
different:

```
array([[15. ,  6. ],
       [ 3.2, -1.5]], dtype=float32)
```

And if you run it yet again, the output is once again different:

```
array([[ 3.2, -1.5],
       [-3. , -2. ]], dtype=float32)
```

And if you run it a fourth time, the output is now:

```
array([[-3., -2.]], dtype=float32)
```

Now that the last record is shown, you have finished streaming
this data. If you run it again, it will return an empty array:

```
array([], shape=(0, 2), dtype=float32)
```

As you can see, the `my_generator` function streams two records in a NumPy array each time it is run. The unique aspect of the generator function is the use of the `yield` statement instead of the `return` statement. Unlike `return`, `yield` produces a sequence of values without storing the entire sequence in the Python runtime memory. `yield` continues to produce a sequence each time we invoke the `next` function until the end of the array is reached.

This example demonstrates how a subset of data can be generated via a generator function. However, in this example, the NumPy array is created on the fly and therefore is held in the Python runtime memory. Let's take a look at how to iterate over a dataset that is stored as a file.

Streaming File Content with a Generator

To understand how a file in the storage can be streamed, you may find it easier to use a CSV file as an example. The file I use here, the Pima Indians Diabetes Dataset, is an open source dataset available for download (*https://oreil.ly/enlwY*). Download it and store it on your local machine.

This file does not contain a header, so you will also need to download (*https://oreil.ly/NxIKk*) the column names and descriptions for this dataset.

Briefly, the columns in this file are:

```
['Pregnancies', 'Glucose', 'BloodPressure',
 'SkinThickness', 'Insulin', 'BMI',
 'DiabetesPedigree', 'Age', 'Outcome']
```

Let's look at this file with the following lines of code:

```
import csv
import pandas as pd

file_path = 'working_data/'
file_name = 'pima-indians-diabetes.data.csv'

col_name = ['Pregnancies', 'Glucose', 'BloodPressure',
            'SkinThickness', 'Insulin', 'BMI',
```

```
                'DiabetesPedigree', 'Age', 'Outcome']
pd.read_csv(file_path + file_name, names = col_name)
```

The first few rows of the file are shown in Figure 2-1.

	Pregnancies	Glucose	BloodPressure	SkinThickness	Insulin	BMI	DiabetesPedigree	Age	Outcome
0	6	148	72	35	0	33.6	0.627	50	1
1	1	85	66	29	0	26.6	0.351	31	0
2	8	183	64	0	0	23.3	0.672	32	1
3	1	89	66	23	94	28.1	0.167	21	0
4	0	137	40	35	168	43.1	2.288	33	1

Figure 2-1. Pima Indians Diabetes Dataset

Since we want to stream this dataset, it is more convenient to read it as a CSV file and use the generator to output the rows, just like we did with the NumPy array in the preceding section. The way to do this is through the following code:

```
import csv
file_path = 'working_data/'
file_name = 'pima-indians-diabetes.data.csv'

with open(file_path + file_name, newline='\n') as csvfile:
    f = csv.reader(csvfile, delimiter=',')
    for row in f:
        print(','.join(row))
```

Let's take a closer look at this code. We use the `with open` command to create a file handle object, `csvfile`, that knows where the file is stored. The next step is to pass it to the `reader` function in the CSV library:

```
f = csv.reader(csvfile, delimiter=',')
```

`f` is the entire file in the Python runtime memory. To inspect the file, execute this short piece of a for loop:

```
for row in f:
        print(','.join(row))
```

The output of the first few rows looks like Figure 2-2.

```
6,148,72,35,0,33.6,0.627,50,1
1,85,66,29,0,26.6,0.351,31,0
8,183,64,0,0,23.3,0.672,32,1
1,89,66,23,94,28.1,0.167,21,0
0,137,40,35,168,43.1,2.288,33,1
```

Figure 2-2. Pima Indians Diabetes Dataset CSV output

Now that you understand how to use a file handle, let's refactor the preceding code so that we can use `yield` in a function, effectively making a generator to stream the content of the file:

```
def stream_file(file_handle):
    holder = []
    for row in file_handle:
        holder.append(row.rstrip("\n"))
        yield holder
        holder = []

with open(file_path + file_name, newline = '\n') as handle:
    for part in stream_file(handle):
        print(part)
```

Recall that a Python generator is a function that uses `yield` to iterate through an `iterable` object. You can use `with open` to acquire a file handle as usual. Then we pass `handle` to a generator function `stream_file`, which contains a `for` loop that iterates through the file in `handle` row by row, removes newline code \n, then fills up a `holder`. Each row is passed back to the main thread's `print` function by `yield` from the generator. The output is shown in Figure 2-3.

```
['6,148,72,35,0,33.6,0.627,50,1']
['1,85,66,29,0,26.6,0.351,31,0']
['8,183,64,0,0,23.3,0.672,32,1']
['1,89,66,23,94,28.1,0.167,21,0']
['0,137,40,35,168,43.1,2.288,33,1']
```

Figure 2-3. Pima Indians Diabetes Dataset output by Python generator

Now that you have a clear idea of how a dataset can be streamed, let's look at how to apply this in TensorFlow. As it turns out, TensorFlow leverages this approach to build a framework for data ingestion. Streaming is usually the best way to ingest large amounts of data (such as hundreds of thousands of rows in one table, or distributed across multiple tables).

JSON Data Structures

Tabular data is a common and convenient format for encoding features and labels for ML model training, and CSV is probably the most common tabular data format. You can think of each field separated by the comma delimiter as a column. Each column is defined with a data type, such as numeric (integer or floating point) or string.

Tabular data is not the only data format that is well structured, by which I mean that every record follows the same convention and the order of fields in every record is the same. Another common data structure is JSON. JSON (JavaScript Object Notation) is a structure built with nested, hierarchical key-value pairs. You can think of keys as column names and values as the actual value of the data in that sample. JSON can be converted to CSV, and vice versa. Sometimes the original data is in JSON format and it is necessary to convert it to CSV, which is easier to display and inspect.

Here's an example JSON record, showing the key-value pairs:

```
{
    "id": 1,
    "name": {
        "first": "Dan",
        "last": "Jones"
    },
    "rating": [
        8,
        7,
        9
    ]
},
```

Notice that the key "rating" is associated with the value of an array [8, 7, 9].

There are plenty of examples of using a CSV file or a table as training data and ingesting it into the TensorFlow model training process. Typically, the data is read into a pandas Data-Frame. However, this strategy only works if all the data can fit into the Python runtime memory. You can use streaming to handle data without the Python runtime restricting memory allocation. Since you learned how a Python generator works in the preceding section, you're now ready to take a look at TensorFlow's API, which operates on the same principle as a Python generator, and learn how to use TensorFlow's adoption of the Python generator framework.

Setting Up a Pattern for Filenames

When working with a set of files, you will encounter patterns in file-naming conventions. To simulate an enterprise environment where new data is continuously being generated and stored, we will use an open source CSV file, split it into multiple parts by row count, then rename each part with a fixed prefix. This approach is similar to how the Hadoop Distributed File System (HDFS) names the parts of a file.

Feel free to use your own CSV file if you have one handy. If not, you can download the suggested CSV file (*https://oreil.ly/8uGKL*) (a COVID-19 dataset) for this example. (You may clone this repository if you wish.)

For now, all you need is *owid-covid-data.csv*. Once it is downloaded, inspect the file and determine the number of rows:

```
wc -l owid-covid-data.csv
```

The output indicates there are over 32,000 rows:

```
32788 owid-covid-data.csv
```

Next, inspect the first three lines of the CSV file to see if there is a header:

```
head -3 owid-covid-data.csv
iso_code,continent,location,date,total_cases,new_cases,
total_deaths,new_deaths,total_cases_per_million,
```

```
new_cases_per_million,total_deaths_per_million,
new_deaths_per_million,new_tests,total_tests,
total_tests_per_thousand,new_tests_per_thousand,
new_tests_smoothed,new_tests_smoothed_per_thousand,tests_units,
stringency_index,population,population_density,median_age,
aged_65_older,aged_70_older,gdp_per_capita,extreme_poverty,
cardiovasc_death_rate,diabetes_prevalence,female_smokers,
male_smokers,handwashing_facilities,hospital_beds_per_thousand,
life_expectancy
AFG,Asia,Afghanistan,2019-12-31,0.0,0.0,0.0,0.0,0.0,0.0,0.0,0.0,
0.0,,,,,,,,,38928341.0,
54.422,18.6,2.581,1.337,1803.987,,597.029,9.59,,,37.746,0.5,64.8
```

Since this file contains a header, you'll see the header in each of the part files. You can also look at a few rows of data to see what they actually look like.

Splitting a Single CSV File into Multiple CSV Files

Now let's split this file into multiple CSV files, each with 330 rows. You should end up with 100 CSV files, each of which has the header. If you use Linux or macOS, use the following command:

```
cat owid-covid-data.csv| parallel --header : --pipe -N330
'cat >owid-covid-data-
part00{#}.csv'
```

For macOS, you may need to first install the `parallel` command:

```
brew install parallel
```

Here are some of the files that are created:

```
-rw-r--r--  1 mbp16  staff   54026 Jul 26 16:45
owid-covid-data-part0096.csv
-rw-r--r--  1 mbp16  staff   54246 Jul 26 16:45
owid-covid-data-part0097.csv
-rw-r--r--  1 mbp16  staff   51278 Jul 26 16:45
owid-covid-data-part0098.csv
-rw-r--r--  1 mbp16  staff   62622 Jul 26 16:45
owid-covid-data-part0099.csv
-rw-r--r--  1 mbp16  staff   15320 Jul 26 16:45
owid-covid-data-part00100.csv
```

This pattern represents a standard storage arrangement for multiple CSV formats. There is a distinct pattern to the naming convention: either all files have the same header, or none has any header at all.

It's a good idea to maintain a file-naming pattern, which can come in handy whether you have tens or hundreds of files. And when your naming pattern can be easily represented with wildcard notation, it's easier to create a reference or file pattern object that points to all the data in storage.

In the next section, we will look at how to use the TensorFlow API to create a file pattern object, which we'll use to create a streaming object for this dataset.

Creating a File Pattern Object Using tf.io

The TensorFlow `tf.io` API is used for referencing a distributed dataset that contains files with a common naming pattern. This is not to say that you want to read the distributed dataset: what you want is a list of file paths and names for all the dataset files you want to read. This is not a new idea. For example, in Python, the glob library (*https://oreil.ly/JLtGm*) is a popular choice for retrieving a similar list. The `tf.io` API simply leverages the glob library to generate a list of filenames that fit the pattern object:

```
import tensorflow as tf

base_pattern = 'dataset'
file_pattern = 'owid-covid-data-part*'
files = tf.io.gfile.glob(base_pattern + '/' + file_pattern)
```

`files` is a list that contains all the CSV filenames that are part of the original CSV, in no particular order:

```
['dataset/owid-covid-data-part0091.csv',
 'dataset/owid-covid-data-part0085.csv',
 'dataset/owid-covid-data-part0052.csv',
 'dataset/owid-covid-data-part0046.csv',
 'dataset/owid-covid-data-part0047.csv',
 ...]
```

This list will be the input for the next step, which is to create a streaming dataset object based on Python generators.

Creating a Streaming Dataset Object

Now that you have your file list ready, you can use it as the input to create a streaming dataset object. Note that this code is *only* meant to demonstrate how to convert a list of CSV files into a TensorFlow dataset object. If you were really going to use this data to train a supervised ML model, you would also perform data cleansing, normalization, and aggregation, all of which we'll cover in Chapter 8. For the purposes of this example, "new_deaths" is selected as the target column:

```
csv_dataset = tf.data.experimental.make_csv_dataset(files,
                header = True,
                batch_size = 5,
                label_name = 'new_deaths',
                num_epochs = 1,
                ignore_errors = True)
```

The preceding code specifies that each file in `files` contains a header. For convenience, as we inspect it, we set a small batch size of 5. We also designate a target column with `label_name`, as if we are going to use this data for training a supervised ML model. `num_epochs` is used to specify how many times you want to stream over the entire dataset.

To look at actual data, you'll need to use the `csv_dataset` object to iterate through the data:

```
for features, target in csv_dataset.take(1):
    print("'Target': {}".format(target))
    print("'Features:'")
    for k, v in features.items():
        print("  {!r:20s}: {}".format(k, v))
```

This code uses the first batch of the dataset (`take(1)`), which contains five samples.

Since you specified `label_name` to be the target column, the other columns are all considered to be features. In the dataset, contents are formatted as key-value pairs. The output from the preceding code will be similar to this:

```
'Target': [ 0.  0. 16.  0.  0.]
'Features:'
  'iso_code'          : [b'SWZ' b'ESP' b'ECU' b'ISL' b'FRO']
  'continent'         :
[b'Africa' b'Europe' b'South America' b'Europe' b'Europe']
  'location'          :
[b'Swaziland' b'Spain' b'Ecuador' b'Iceland' b'Faeroe Islands']
  'date'              :
[b'2020-04-04' b'2020-02-07' b'2020-07-13' b'2020-04-01'
  b'2020-06-11']
  'total_cases'       : [9.000e+00 1.000e+00 6.787e+04
1.135e+03 1.870e+02]
  'new_cases'         : [  0.    0. 661.  49.   0.]
  'total_deaths'      : [0.000e+00 0.000e+00 5.047e+03
2.000e+00 0.000e+00]
  'total_cases_per_million':
              [7.758000e+00 2.100000e-02 3.846838e+03
3.326007e+03 3.826870e+03]
  'new_cases_per_million': [  0.      0.     37.465
143.59    0.   ]
  'total_deaths_per_million': [  0.      0.    286.061
5.861    0.   ]
  'new_deaths_per_million':
[0.    0.    0.907 0.    0.    ]
  'new_tests'         :
[b'' b'' b'1331.0' b'1414.0' b'']
  'total_tests'       :
[b'' b'' b'140602.0' b'20889.0' b'']
  'total_tests_per_thousand':
[b'' b'' b'7.969' b'61.213' b'']
  'new_tests_per_thousand':
[b'' b'' b'0.075' b'4.144' b'']
  'new_tests_smoothed':
[b'' b'' b'1986.0' b'1188.0' b'']
  'new_tests_smoothed_per_thousand':
[b'' b'' b'0.113' b'3.481' b'']
  'tests_units'       :
[b'' b'' b'units unclear' b'tests performed' b'']
  'stringency_index'  :
[89.81 11.11 82.41 53.7   0.  ]
  'population'        :
[ 1160164. 46754784. 17643060.   341250.    48865.]
  'population_density':
```

```
[79.492 93.105 66.939  3.404 35.308]
  'median_age'        :
[21.5 45.5 28.1 37.3  0. ]
  'aged_65_older'     :
[ 3.163 19.436  7.104 14.431  0.    ]
  'aged_70_older'     :
[ 1.845 13.799  4.458  9.207  0.    ]
  'gdp_per_capita'    :
[ 7738.975 34272.36  10581.936 46482.957     0.    ]
  'extreme_poverty'   : [b'' b'1.0' b'3.6' b'0.2' b'']
  'cardiovasc_death_rate':
[333.436  99.403 140.448 117.992   0.    ]
  'diabetes_prevalence': [3.94 7.17 5.55 5.31 0.  ]
  'female_smokers'    :
[b'1.7' b'27.4' b'2.0' b'14.3' b'']
  'male_smokers'      :
[b'16.5' b'31.4' b'12.3' b'15.2' b'']
  'handwashing_facilities':
[24.097  0.     80.635  0.      0.    ]
  'hospital_beds_per_thousand':
[2.1  2.97 1.5  2.91 0.  ]
  'life_expectancy'   :
[60.19 83.56 77.01 82.99 80.67]
```

This data is retrieved during runtime (lazy execution). As indicated by the batch size, each column contains five records. Next, let's discuss how to stream this dataset.

Streaming a CSV Dataset

Now that a CSV dataset object has been created, you can easily iterate over it in batches with this line of code, which uses the `iter` function to make an iterator from the CSV dataset and the `next` function to return the next item in the iterator:

```
features, label = next(iter(csv_dataset))
```

Remember that in this dataset there are two types of elements: `features` and `label`. These elements are returned as a *tuple* (similar to a list of objects, except that the order and the value of objects cannot be changed or reassigned). You can unpack a tuple by assigning the tuple elements to variables.

If you examine the label, you'll see the content of the first batch:

```
<tf.Tensor: shape=(5,), dtype=float32,
numpy=array([ 0.,  0.,  1., 33., 29.], dtype=float32)>
```

If you execute the same command again, you'll see the second batch:

```
features, label = next(iter(csv_dataset))
```

Let's just take a look at label:

```
<tf.Tensor: shape=(5,), dtype=float32,
numpy=array([ 7., 15.,  1.,  0.,  6.], dtype=float32)>
```

Indeed, this is the second batch of observations; it contains different values than the first batch. This is how a streaming CSV dataset is produced in a data ingestion pipeline. As each batch is sent to the model for training, the model computes the prediction in the *forward pass*, which computes the output by multiplying the input value and the current weight and bias in each node of the neural network. Then it compares the prediction with the label and calculates the loss function. Next comes the *backward pass*, where the model computes the variation with respect to the expected output and goes backward into each node of the network to update the weight and bias. The model then recalculates and updates the gradients. A new batch of data is sent to the model for training, and the process repeats.

Next we will look at how to organize image data for storage and stream it like we streamed the structured data.

Organizing Image Data

Image classification tasks require organizing images in certain ways because, unlike CSV or tabular data, attaching a label to an image requires special techniques. A straightforward and common pattern for organizing image files is with the following directory structure:

```
<PROJECT_NAME>
    train
        class_1
            <FILENAME>.jpg
            <FILENAME>.jpg
            ...
        class_n
            <FILENAME>.jpg
            <FILENAME>.jpg
            ...
    validation
        class_1
            <FILENAME>.jpg
            <FILENAME>.jpg
            ...
        class_n
            <FILENAME>.jpg
            <FILENAME>.jpg
            ...
    test
        class_1
            <FILENAME>.jpg
            <FILENAME>.jpg
            ...
        class_n
            <FILENAME>.jpg
            <FILENAME>.jpg
            ...
```

<PROJECT_NAME> is the base directory. The first level below
it contains training, validation, and test directories. Within
each of these directories, there are subdirectories named with
the image labels (class_1, class_2, etc., which in the following
example are flower types), each of which contains the raw
image files. This is shown in Figure 2-4.

This structure is common because it makes it easy to keep track
of labels and their respective images, but by no means is it the
only way to organize image data. Let's look at another structure
for organizing images. This is very similar to the previous one,
except that training, testing, and validation are all separate.
Immediately below the <PROJECT_NAME> directory are the
directories of different image classes, as shown in Figure 2-5.

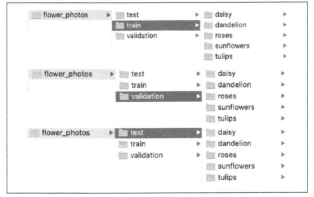

Figure 2-4. File organization for image classification and partitioning for training work

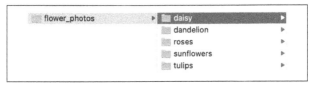

Figure 2-5. File organization for images based on labels

Using TensorFlow Image Generator

Now let's take a look at how to deal with images. Besides the nuances of file organization, working with images also requires certain steps to standardize and normalize the image files. The model architecture requires a fixed shape (fixed dimensions) for all images. At the pixel level, the values are normalized, typically to a range of [0, 1] (dividing the pixel value by 255).

For this example, you'll use an open source image set of five different types of flowers (or feel free to use your own image set). Let's assume that images should be 224 × 224 pixels, where the dimensions correspond to height and width. These are the expected dimensions for input images if you want to use a pretrained residual neural network (ResNet) as the image classifier.

First let's download the images. The following code downloads five types of flowers, all in different dimensions, and puts them in the file structure shown later in Figure 2-6:

```
import tensorflow as tf

data_dir = tf.keras.utils.get_file(
    'flower_photos',
'https://storage.googleapis.com/download.tensorflow.org/
example_images/flower_photos.tgz', untar=True)
```

We will refer to `data_dir` as the base directory. It should be similar to:

```
'/Users/XXXXX/.keras/datasets/flower_photos'
```

If you list the content from the base directory, you'll see:

```
-rw-r-----   1 mbp16  staff  418049 Feb  8  2016 LICENSE.txt
drwx------ 801 mbp16  staff   25632 Feb 10  2016 tulips
drwx------ 701 mbp16  staff   22432 Feb 10  2016 sunflowers
drwx------ 643 mbp16  staff   20576 Feb 10  2016 roses
drwx------ 900 mbp16  staff   28800 Feb 10  2016 dandelion
drwx------ 635 mbp16  staff   20320 Feb 10  2016 daisy
```

There are three steps to streaming the images. Let's look more closely:

1. Create an `ImageDataGenerator` object and specify normalization parameters. Use the `rescale` parameter to indicate the normalization scale and the `validation_split` parameter to specify that 20% of the data will be set aside for cross validation:

   ```
   train_datagen = tf.keras.preprocessing.image.
       ImageDataGenerator(
       rescale = 1./255,
       validation_split = 0.20)
   ```

   Optionally, you can wrap `rescale` and `validation_split` as a dictionary that consists of key-value pairs:

   ```
   datagen_kwargs = dict(rescale=1./255,
                         validation_split=0.20)
   ```

```
train_datagen = tf.keras.preprocessing.image.
    ImageDataGenerator(**datagen_kwargs)
```

This is a convenient way to reuse the same parameters and keep multiple input arguments under wrap. (Passing the dictionary data structure to a function is a Python technique known as *dictionary unpacking*.)

2. Connect the ImageDataGenerator object to the data source and specify parameters to resize the images to a fixed dimension:

```
IMAGE_SIZE = (224, 224) # Image height and width
BATCH_SIZE = 32
dataflow_kwargs = dict(target_size=IMAGE_SIZE,
                       batch_size=BATCH_SIZE,
                       interpolation="bilinear")

train_generator = train_datagen.flow_from_directory(
data_dir, subset="training", shuffle=True,
**dataflow_kwargs)
```

3. Prepare a map for indexing the labels. In this step, you retrieve the index that the generator has assigned to each label and create a dictionary that maps it to the actual label name. The TensorFlow generator internally keeps track of labels from the directory name below data_dir. They can be retrieved through train_generator.class_indices, which returns a key-value pair of labels and indices. You can take advantage of this and reverse it to deploy the model for scoring. The model will output the index. To implement this reverse lookup, simply reverse the label dictionary generated by train_generator.class_indices:

```
labels_idx = (train_generator.class_indices)
idx_labels = dict((v,k) for k,v in labels_idx.items())
```

These are the idx_labels:

```
{0: 'daisy', 1: 'dandelion', 2: 'roses',
  3: 'sunflowers', 4: 'tulips'}
```

Now you can inspect the shape of the items generated by `train_generator`:

```
for image_batch, labels_batch in train_generator:
  print(image_batch.shape)
  print(labels_batch.shape)
  break
```

Expect to see the following for the first batch yielded by the generator iterating through the base directory:

```
(32, 224, 224, 3)
(32, 5)
```

The first tuple indicates a batch size of 32 images, each with a dimension of 224 × 224 × 3 (height × width × depth, where depth represents the three color channels RGB). The second tuple indicates 32 labels, each corresponding to one of the five flower types. It is one-hot encoded per `idx_labels`.

Streaming Cross-Validation Images

Recall that in creating the generator for streaming training data, you specified the `validation_split` parameter with a value of 0.2. If you don't do this, `validation_split` defaults to a value of 0. If `validation_split` is set to a nonzero decimal, when you invoke the `flow_from_directory` method, you also have to specify `subset` to be either `training` or `validation`. In the preceding example, it is `subset="training"`.

You may be wondering how you'll know which images belong to the `training` subset from our previous endeavor of creating a training generator. Well, you don't have to know this if you reassign and reuse the training generator:

```
valid_datagen = train_datagen

valid_generator = valid_datagen.flow_from_directory(
    data_dir, subset="validation", shuffle=False,
    **dataflow_kwargs)
```

As you can see, a TensorFlow generator knows and keeps track of training and validation subsets, so you can reuse the same generator to stream over different subsets. The dataflow_kwargs dictionary is also reused. This is a convenience feature provided by TensorFlow generators.

Because you reuse train_datagen, you can be sure that image rescaling is done the same way as image training. And in the valid_datagen.flow_from_directory method, you'll pass in the same dataflow_kwargs dictionary to set the image size for cross validation to be the same as it is for the training images.

If you prefer to organize the images into training, validation, and testing yourself, what you learned earlier still applies, with two exceptions. First, your data_dir is at the level of the training, validation, or testing directory. Second, you don't need to specify validation_split in ImageDataGenerator and subset in flow_from_directory.

Inspecting Resized Images

Now let's inspect the resized images coming off the generator. Following is the code snippet for iterating through a batch of data streamed by a generator:

```
import matplotlib.pyplot as plt
import numpy as np

image_batch, label_batch = next(iter(train_generator))

fig, axes = plt.subplots(8, 4, figsize=(10, 20))
axes = axes.flatten()
for img, lbl, ax in zip(image_batch, label_batch, axes):
    ax.imshow(img)
    label_ = np.argmax(lbl)
    label = idx_labels[label_]
    ax.set_title(label)
    ax.axis('off')
plt.show()
```

This code will produce 32 images from the first batch coming off the generator (see Figure 2-6).

Figure 2-6. A batch of reshaped images

Let's examine the code:

```
image_batch, label_batch = next(iter(train_generator))
```

This iterates over the base directory with the generator. It applies the `iter` function to the generator and leverages the `next` function to output the image batch and label batch as NumPy arrays:

```
fig, axes = plt.subplots(8, 4, figsize=(10, 20))
```

This line sets up the number of subplots you expect, which is 32, your batch size:

```
axes = axes.flatten()
for img, lbl, ax in zip(image_batch, label_batch, axes):
    ax.imshow(img)
    label_ = np.argmax(lbl)
    label = idx_labels[label_]
    ax.set_title(label)
    ax.axis('off')
plt.show()
```

Then you set the figure axes, using a for loop to display NumPy arrays as images and labels. As shown in Figure 2-6, all the images are resized into 224 × 224-pixel squares. Although the subplot holder is a rectangle with `figsize=(10, 20)`, you can see that all of the images are squares. This means your code for resizing and normalizing images in the generator workflow works as expected.

Wrapping Up

In this chapter, you learned the fundamentals of streaming data using Python. This is a workhorse technique when working with large, distributed datasets. You also saw some common file organization patterns for tabular and image data.

In the section on tabular data, you learned how choosing a good file-naming convention can make it easier to build a reference to all the files, regardless of how many there are. This means you now know how to build a scalable pipeline that can ingest as much data as needed into a Python runtime for any use (in this case, for TensorFlow to create a dataset).

You also learned how image files are usually organized in file storage and how to associate images with labels. In the next chapter, you will leverage what you've learned here about data organization and streaming to integrate it with the model training process.

## CHAPTER 3

Data Preprocessing

In this chapter, you'll learn how to prepare and set up data for training. Some of the most common data formats for ML work are tables, images, and text. There are commonly practiced techniques associated with each, though how you set up your data engineering pipeline will, of course, depend on what your problem statement is and what you are trying to predict.

I'll look at all three formats in detail, using specific examples to walk you through the techniques. All the data can be read directly into your Python runtime memory; however, this isn't the most efficient way to use your compute resources. When I discuss text data, I'll give particular attention to tokenization and dictionaries. By the end of this chapter, you'll have learned how to prepare table, image, and text data for training.

Preparing Tabular Data for Training

In a tabular dataset, it is important to identify which columns are considered categorical, because you have to encode their value as a class or a binary representation of the class (one-hot encoding), rather than a numerical value. Another aspect of tabular datasets is the potential for interactions among multiple features. This section will also look at the API that TensorFlow provides to make it easier to model column interactions.

It's common to encounter tabular datasets as CSV files or simply as structured output from a database query. For this example, we'll start with a dataset that's already in a pandas DataFrame and then learn how to transform it and set it up for model training. We'll use the *Titanic* dataset, an open source, tabular dataset that is often used for teaching because of its manageable size and availability. This dataset contains attributes for each passenger, such as age, gender, cabin grade, and whether or not they survived. We are going to try to predict each passenger's probability of survival based on their attributes or features. Be aware that this is a small dataset for teaching and learning purposes only. In reality, your dataset will likely be much larger. You may make different decisions and choose different default values for some of these input parameters, so don't take this example too literally.

Let's start with loading all the necessary libraries:

```
import functools
import numpy as np
import tensorflow as tf
import pandas as pd
from tensorflow import feature_column
from tensorflow.keras import layers
from sklearn.model_selection import train_test_split
```

Load the data from Google's public storage:

```
TRAIN_DATA_URL = "https://storage.googleapis.com/
tf-datasets/titanic/train.csv"
TEST_DATA_URL = "https://storage.googleapis.com/
tf-datasets/titanic/eval.csv"

train_file_path = tf.keras.utils.get_file("train.csv",
TRAIN_DATA_URL)
test_file_path = tf.keras.utils.get_file("eval.csv", TEST_DATA_URL)
```

Now take a look at `train_file_path`:

```
print(train_file_path)
```

```
/root/.keras/datasets/train.csv
```

This file path points to a CSV file, which we'll read as a pandas DataFrame:

```
titanic_df = pd.read_csv(train_file_path, header='infer')
```

Figure 3-1 shows what `titanic_df` looks like as a pandas DataFrame.

	survived	sex	age	n_siblings_spouses	parch	fare	class	deck	embark_town	alone
0	0	male	22.0	1	0	7.2500	Third	unknown	Southampton	n
1	1	female	38.0	1	0	71.2833	First	C	Cherbourg	n
2	1	female	26.0	0	0	7.9250	Third	unknown	Southampton	y
3	1	female	35.0	1	0	53.1000	First	C	Southampton	n
4	0	male	28.0	0	0	8.4583	Third	unknown	Queenstown	y
...	...	...	...	...	...	...	...	...	...	...
622	0	male	28.0	0	0	10.5000	Second	unknown	Southampton	y
623	0	male	25.0	0	0	7.0500	Third	unknown	Southampton	y
624	1	female	19.0	0	0	30.0000	First	B	Southampton	y
625	0	female	28.0	1	2	23.4500	Third	unknown	Southampton	n
626	0	male	32.0	0	0	7.7500	Third	unknown	Queenstown	y

627 rows × 10 columns

Figure 3-1. Titanic dataset as a pandas DataFrame

Marking Columns

As you can see in Figure 3-1, there are numeric as well as categorical columns in this data. The target column, or the column for prediction, is the "survived" column. You'll need to mark it as the target and mark the rest of the columns as features.

TIP

A best practice in TensorFlow is to convert your table into a streaming dataset. This practice ensures that the data's size does not affect memory consumption.

To do exactly that, TensorFlow provides the function `tf.data.experimental.make_csv_dataset`:

```
LABEL_COLUMN = 'survived'
LABELS = [0, 1]

train_ds = tf.data.experimental.make_csv_dataset(
    train_file_path,
    batch_size=3,
    label_name=LABEL_COLUMN,
```

```
        na_value="?",
        num_epochs=1,
        ignore_errors=True)

test_ds = tf.data.experimental.make_csv_dataset(
        test_file_path,
        batch_size=3,
        label_name=LABEL_COLUMN,
        na_value="?",
        num_epochs=1,
        ignore_errors=True)
```

In the preceding function signature, you specify the file path
for which you wish to generate a dataset object. The `batch_size`
is arbitrarily set to something small (3 in this case) for convenience in inspecting the data. We also set `label_name` to the
"survived" column. For data quality, if a question mark (?) is
specified in any cell, you want it to be interpreted as "NA" (not
applicable). For training, set `num_epochs` to iterate over the
dataset once. You can ignore any parsing errors or empty lines.

Next, inspect the data:

```
for batch, label in train_ds.take(1):
  print(label)
  for key, value in batch.items():
    print("{}: {}".format(key,value.numpy()))
```

It will appear similar to Figure 3-2.

```
    tf.Tensor([1 1 0], shape=(3,), dtype=int32)
    sex: [b'female' b'female' b'male']
    age: [51. 30. 28.]
    n_siblings_spouses: [1 3 0]
    parch: [0 0 0]
    fare: [77.958 21.      7.854]
    class: [b'First' b'Second' b'Third']
    deck: [b'D' b'unknown' b'unknown']
    embark_town: [b'Southampton' b'Southampton' b'Southampton']
    alone: [b'n' b'n' b'y']
```

Figure 3-2. A batch of the Titanic dataset

Here are the major steps for training a paradigm to consume your training dataset:

1. Designate columns by feature types.

2. Decide whether or not to embed or cross columns.

3. Choose the columns of interest, possibly as an experiment.

4. Create a "feature layer" for consumption by the training paradigm.

Now that you have set up the data as datasets, you can designate each column by its feature type, such as numeric or categorical, bucketized (by binning) if necessary. You can also embed the column if there are too many unique categories and dimension reduction would be helpful.

Let's go ahead with step 1. There are four numeric columns: age, n_siblings_spouses, parch, and fare. Five columns are categorical: sex, class, deck, embark_town, and alone. You will create a feature_columns list to hold all the feature columns once you are done.

Here is how to designate numeric columns based strictly on the actual numeric values, without any transformation:

```
feature_columns = []

# numeric cols
for header in ['age', 'n_siblings_spouses', 'parch', 'fare']:
feature_columns.append(feature_column.numeric_column(header))
```

Note that in addition to using age as is, you could also bin age into a bucket, such as by quantile of age distribution. But what are the bin boundaries (quantiles)? You can inspect the general statistics of numeric columns in a pandas DataFrame:

```
titanic_df.describe()
```

Figure 3-3 shows the output.

	survived	age	n_siblings_spouses	parch	fare
count	627.000000	627.000000	627.000000	627.000000	627.000000
mean	0.387560	29.631308	0.545455	0.379585	34.385399
std	0.487582	12.511818	1.151090	0.792999	54.597730
min	0.000000	0.750000	0.000000	0.000000	0.000000
25%	0.000000	23.000000	0.000000	0.000000	7.895800
50%	0.000000	28.000000	0.000000	0.000000	15.045800
75%	1.000000	35.000000	1.000000	0.000000	31.387500
max	1.000000	80.000000	8.000000	5.000000	512.329200

Figure 3-3. Statistics for numeric columns in the Titanic dataset

Let's try three bin boundaries for age: 23, 28, and 35. This means passenger age will be grouped into first quantile, second quantile, and third quantile (as shown in Figure 3-3):

```
age = feature_column.numeric_column('age')
age_buckets = feature_column.
bucketized_column(age, boundaries=[23, 28, 35])
```

Therefore, in addition to "age," you have generated another column, "age_bucket."

To understand the nature of each categorical column, it would be helpful to know the distinct values in them. You'll need to encode the vocabulary list with the unique entries in each column. For categorical columns, this means you need to determine which entries are unique:

```
h = {}
for col in titanic_df:
  if col in ['sex', 'class', 'deck', 'embark_town', 'alone']:
    print(col, ':', titanic_df[col].unique())
    h[col] = titanic_df[col].unique()
```

The result is shown in Figure 3-4.

```
sex : ['male' 'female']
class : ['Third' 'First' 'Second']
deck : ['unknown' 'C' 'G' 'A' 'B' 'D' 'F' 'E']
embark_town : ['Southampton' 'Cherbourg' 'Queenstown' 'unknown']
alone : ['n' 'y']
```

Figure 3-4. Unique values in each categorical column of the dataset

You need to keep track of these unique values in a dictionary format for the model to do the mapping and lookup. Therefore, you'll encode unique categorical values in the "sex" column:

```
sex_type = feature_column.categorical_column_with_vocabulary_list(
    'Type', ['male' 'female'])
sex_type_one_hot = feature_column.indicator_column(sex_type)
```

However, if the list is long, it becomes inconvenient to write it out. Instead, as you iterate through the categorical columns, you can save each column's unique values in a Python dictionary data structure h for future lookup. Then you can pass the unique value as a list into these vocabulary lists:

```
sex_type = feature_column.
categorical_column_with_vocabulary_list(
    'Type', h.get('sex').tolist())
sex_type_one_hot = feature_column.
indicator_column(sex_type)

class_type = feature_column.
categorical_column_with_vocabulary_list(
    'Type', h.get('class').tolist())
class_type_one_hot = feature_column.
indicator_column(class_type)

deck_type = feature_column.
categorical_column_with_vocabulary_list(
    'Type', h.get('deck').tolist())
deck_type_one_hot = feature_column.
indicator_column(deck_type)

embark_town_type = feature_column.
categorical_column_with_vocabulary_list(
    'Type', h.get('embark_town').tolist())
embark_town_type_one_hot = feature_column.
indicator_column(embark_town_type)

alone_type = feature_column.
categorical_column_with_vocabulary_list(
    'Type', h.get('alone').tolist())
alone_one_hot = feature_column.
indicator_column(alone_type)
```

You can also embed the "deck" column, since there are eight unique values, more than any other categorical column. Reduce its dimension to 3:

```
deck = feature_column.
categorical_column_with_vocabulary_list(
    'deck', titanic_df.unique())
deck_embedding = feature_column.
embedding_column(deck, dimension=3)
```

Another way to reduce the dimensions of categorical columns is by using a *hashed feature column*. This method calculates a hashed value based on the input data. It then designates a hashed bucket for the data. The following code reduces the dimension of the "class" column to 4:

```
class_hashed = feature_column.categorical_column_with_hash_bucket(
    'class', hash_bucket_size=4)
```

Encoding Column Interactions as Possible Features

Now comes the most interesting part: you're going to find interactions between different features (this is referred to as *crossing columns*) and encode those interactions as possible features. This is also where your intuition and domain knowledge can benefit your feature engineering endeavor. For example, a question that comes to mind based on the historical background of the *Titanic* disaster is this: were women in first-class cabins more likely to survive than women in second- or third-class cabins? To rephrase this as a data science question, you'll need to consider interactions between the gender and cabin class of the passengers. Then you'll need to pick a starting dimension size to represent the data variability. Let's say you arbitrarily decide to bin the variability into five dimensions (hash_bucket_size):

```
cross_type_feature = feature_column.
crossed_column(['sex', 'class'], hash_bucket_size=5)
```

Now that you have created all the features, you need to put them together—and perhaps experiment to decide which to include in the training process. For that, you'll first create a list to hold all the features you want to use:

```
feature_columns = []
```

Then you'll append each feature of interest to the list:

```
# append numeric columns
for header in ['age', 'n_siblings_spouses', 'parch', 'fare']:
  feature_columns.append(feature_column.numeric_column(header))

# append bucketized columns
age = feature_column.numeric_column('age')
age_buckets = feature_column.
bucketized_column(age, boundaries=[23, 28, 35])
feature_columns.append(age_buckets)

# append categorical columns
indicator_column_names =
['sex', 'class', 'deck', 'embark_town', 'alone']
for col_name in indicator_column_names:
  categorical_column = feature_column.
  categorical_column_with_vocabulary_list(
      col_name, titanic_df[col_name].unique())
  indicator_column = feature_column.

indicator_column(categorical_column)
  feature_columns.append(indicator_column)

# append embedding columns
deck = feature_column.categorical_column_with_vocabulary_list(
      'deck', titanic_df.deck.unique())
deck_embedding = feature_column.
embedding_column(deck, dimension=3)
feature_columns.append(deck_embedding)

# append crossed columns
feature_columns.
append(feature_column.indicator_column(cross_type_feature))
```

Now create a feature layer:

```
feature_layer = tf.keras.layers.DenseFeatures(feature_columns)
```

This layer will serve as the first (input) layer in the model you are about to build and train. This is how you'll provide all the feature engineering frameworks for the model's training process.

Creating a Cross-Validation Dataset

Before you start training, you have to create a small dataset for cross-validation purposes. Since there are only two partitions (training and testing) to begin with, one way to generate a cross-validation dataset is to simply subdivide one of the partitions:

```
val_df, test_df = train_test_split(test_df, test_size=0.4)
```

Here, 40% of the original `test_df` partition was randomly reserved as `test_df`, and the remaining 60% is now `val_df`. Usually, test datasets are the smallest of the three (training, validation, testing), since they are used only for final evaluation, and not during model training.

Now that you have taken care of feature engineering and data partitioning, there is one last thing to do: stream the data into the training process with the dataset. You'll convert each of the three DataFrames (training, validation, and testing) into its own dataset:

```
batch_size = 32
labels = train_df.pop('survived')
working_ds = tf.data.Dataset.
from_tensor_slices((dict(train_df), labels))
working_ds = working_ds.shuffle(buffer_size=len(train_df))
train_ds = working_ds.batch(batch_size)
```

As shown in the preceding code, first you'll arbitrarily decide how many samples to include in a batch (`batch_size`). Then you need to set aside a label designation (`survived`). The `tf.data.Dataset.from_tensor_slices` method takes a tuple as an argument. In this tuple, there are two elements: feature columns and the label column.

The first element is `dict(train_df)`. This `dict` operation essentially transforms the DataFrame into a key-value pair, where each key represents a column name and the corresponding value is an array of the values in the column. The other element is `labels`.

Finally, we shuffle and batch the dataset. Since this conversion will be applied to all three datasets, it would be convenient to combine these steps into a helper function to reduce repetition:

```
def pandas_to_dataset(dataframe, shuffle=True, batch_size=32):
  dataframe = dataframe.copy()
  labels = dataframe.pop('survived')
  ds = tf.data.Dataset.
from_tensor_slices((dict(dataframe), labels))
  if shuffle:
    ds = ds.shuffle(buffer_size=len(dataframe))
  ds = ds.batch(batch_size)
  return ds
```

Now you can apply this function to both validation and test data:

```
val_ds = pandas_to_dataset(val_df, shuffle=False,
batch_size=batch_size)
test_ds = pandas_to_dataset(test_df, shuffle=False,
batch_size=batch_size)
```

Starting the Model Training Process

Now you're ready to start the model training process. Technically this isn't a part of preprocessing, but running through this short section will allow you to see how the work you have done fits into the model training process itself.

You'll start by building a model architecture:

```
model = tf.keras.Sequential([
  feature_layer,
  layers.Dense(128, activation='relu'),
  layers.Dense(128, activation='relu'),
  layers.Dropout(.1),
  layers.Dense(1)
])
```

For demonstration purposes, you'll build a simple two-layer, deep-learning perceptron model, which is a basic configuration of a feedforward neural network. (For more on this, see Aurélien Géron's *Neural Networks and Deep Learning* (*https://oreil.ly/1wOQj*) (O'Reilly)). Notice that since this is a multilayer perceptron model, you'll use the sequential API. Inside this API, the first layer is feature_layer, which represents all the feature engineering logic and derived features, such as age bins and crosses, that are used to model the feature interactions.

Compile the model and set up the loss function for binary classification:

```
model.compile(optimizer='adam',
              loss=tf.keras.losses.BinaryCrossentropy(
from_logits=True),
              metrics=['accuracy'])
```

Then you can start the training. You'll only train it for 10 epochs:

```
model.fit(train_ds,
          validation_data=val_ds,
          epochs=10)
```

You can expect an outcome similar to that pictured in Figure 3-5.

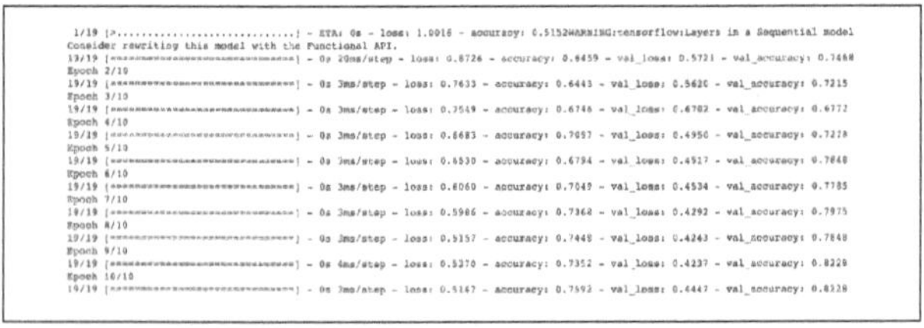

Figure 3-5. Example training outcome from survival prediction in the Titanic dataset

Summary

In this section, you saw how to deal with tabular data that consists of multiple data types. You also saw that TensorFlow provides a `feature_column` API, which enables proper casting of data types, handling of categorical data, and feature crossing for potential interactions. This API is very helpful in simplifying data and feature engineering tasks.

Preparing Image Data for Processing

For images, you need to reshape or resample all the images into the same pixel count; this is known as *standardization*. You also need to ensure that all pixel values are within the same color range so that they fall within the finite range of RGB values of each pixel.

Image data comes with different file extensions, such as *.jpg*, *.tiff*, and *.bmp*. These are not really problematic, as there are APIs in Python and TensorFlow that can read and parse images with any file extension. The tricky part about image data is capturing its dimensions—height, width, and depth—as measured by pixel counts. (If it is a color image encoded with RGB, these appear as three separate channels.)

If all the images in your dataset (including training, validation, and all the images during testing or deployment time) are expected to have the same dimensions *and* you are going to build your own model, then processing image data is not too much of a problem. However, if you wish to leverage prebuilt models such as ResNet or Inception, then you have to conform to their image requirements. As an example, ResNet requires each input image to be $224 \times 224 \times 3$ pixels and be presented as a NumPy multidimensional array. This means that, in the preprocessing routine, you have to resample your images to conform to those dimensions.

Another situation for resampling arises when you cannot reasonably expect all the images, especially during deployment, to have the same size. In this case, you need to consider a proper image dimension as you build the model, then set up the preprocessing routine to ensure that resampling is done properly.

In this section, you are going to use the flower dataset provided by TensorFlow. It consists of five types of flowers and diverse image dimensions. This is a convenient dataset to use, since all images are already in JPEG format. You are going to process this image data to train a model to parse each image and classify it as one of the five classes of flowers.

As usual, import all necessary libraries:

```
import tensorflow as tf
import numpy as np
import matplotlib.pylab as plt
import pathlib
```

Now download the flower dataset from the URL:

```
data_dir = tf.keras.utils.get_file(
    'flower_photos',
'https://storage.googleapis.com/download.tensorflow.org/
example_images/flower_photos.tgz',
    untar=True)
```

This file is a compressed TAR archive file. Therefore, you need to set untar=True.

When using tf.keras.utils.get_file, by default you will find the downloaded data in the ~/.keras/datasets directory.

In a Mac or Linux system's Jupyter Notebook cell, execute:

```
!ls -lrt ~/.keras/datasets/flower_photos
```

You will find the flower dataset as shown in Figure 3-6.

```
total 604
-rw-r----- 1 270850 5000 418049 Feb  9  2016 LICENSE.txt
drwx------ 2 270850 5000  36864 Feb 10  2016 tulips
drwx------ 2 270850 5000  36864 Feb 10  2016 sunflowers
drwx------ 2 270850 5000  36864 Feb 10  2016 roses
drwx------ 2 270850 5000  49152 Feb 10  2016 dandelion
drwx------ 2 270850 5000  36864 Feb 10  2016 daisy
```

Figure 3-6. Flower dataset folders

Now let's take a look at one of the flower types:

```
!ls -lrt ~/.keras/datasets/flower_photos/roses | head -10
```

You should see the first nine images, as shown in Figure 3-7.

```
-rw-r----- 1 270850 5000   33399 Jan 11  2016 7409458444_0bfc9a0682_n.jpg
-rw-r----- 1 270850 5000  102758 Jan 11  2016 7345657862_689366e79a.jpg
-rw-r----- 1 270850 5000   78640 Jan 11  2016 9337528427_3d09b7012b.jpg
-rw-r----- 1 270850 5000   32020 Jan 11  2016 7551637034_55ae047756_n.jpg
-rw-r----- 1 270850 5000  108384 Jan 11  2016 5736328472_8f25e6f6e7.jpg
-rw-r----- 1 270850 5000   40554 Jan 11  2016 3997609936_8db20b7141_n.jpg
-rw-r----- 1 270850 5000   24761 Jan 11  2016 3903276582_fe05bf84c7_n.jpg
-rw-r----- 1 270850 5000   39699 Jan 11  2016 3751835302_d5a03f55e8_n.jpg
-rw-r----- 1 270850 5000   26655 Jan 11  2016 3667366832_7a8017c528_n.jpg
```

Figure 3-7. Nine example image files in the rose directory

These images are all different sizes. You can verify this by examining a couple of images. Here is a helper function you can leverage to display the image in its original size:[1]

```python
def display_image_in_actual_size(im_path):

    dpi = 100
    im_data = plt.imread(im_path)
    height, width, depth = im_data.shape
    # What size does the figure need to be in inches to fit
    # the image?
    figsize = width / float(dpi), height / float(dpi)
    # Create a figure of the right size with one axis that
    # takes up the full figure
    fig = plt.figure(figsize=figsize)
    ax = fig.add_axes([0, 0, 1, 1])
    # Hide spines, ticks, etc.
    ax.axis('off')
    # Display the image.
    ax.imshow(im_data, cmap='gray')
    plt.show()
```

Let's use it to display an image (as shown in Figure 3-8):

```python
IMAGE_PATH = "/root/.keras/datasets/flower_photos/roses/
7409458444_0bfc9a0682_n.jpg"
display_image_in_actual_size(IMAGE_PATH)
```

---

1 From an answer on StackOverflow (*https://oreil.ly/1iVv1*) by user Joe Kington, January 13, 2016, accessed October 23, 2020.

Figure 3-8. Rose image sample 1

Now try a different image (as shown in Figure 3-9):

```
IMAGE_PATH = "/root/.keras/datasets/flower_photos/roses/
5736328472_8f25e6f6e7.jpg"
display_image_in_actual_size(IMAGE_PATH)
```

Figure 3-9. Rose image sample 2

Clearly, the dimensions and aspect ratios of these images are different.

Transforming Images to a Fixed Specification

Now you're ready to transform these images to a fixed specification. In this particular example, you'll use the ResNet input image spec, which is 224 × 224 with three color channels (RGB). Also, it is a best practice to use data streaming whenever possible. Therefore, your goal here is to transform these color images into the shape of 224 × 224 pixels and build a dataset from them for streaming into the training paradigm.

To accomplish this you'll use the ImageDataGenerator class and the flow_from_directory method.

ImageDataGenerator is responsible for creating a generator object, which generates streaming data from the directory as specified by flow_from_directory.

In general, the coding pattern is:

```
my_datagen = tf.keras.preprocessing.image.ImageDataGenerator(
    **datagen_kwargs)
my_generator = my_datagen.flow_from_directory(
data_dir, **dataflow_kwargs)
```

In both cases, keyword argument options, or kwargs, give your code great flexibility. (Keyword arguments are frequently seen in Python.) These arguments enable you to pass optional parameters to the function. As it turns out, in ImageDataGenerator, there are two parameters relevant to your needs: rescale and validation_split. The rescale parameter is used for normalizing pixel values into a finite range, and validation_split lets you subdivide a partition of data, such as for cross validation.

In flow_from_directory, there are three parameters that are useful for this example: target_size, batch_size, and interpolation. The target_size parameter helps you specify the desired dimension of each image, and batch_size is for specifying the number of samples in a batch of images. As for interpolation, remember how you need to interpolate, or resample, each image to a prescribed dimension specified with target_size? Supported methods for interpolation are nearest, bilinear, and bicubic. For this example, first try bilinear.

You can define these keyword arguments as follows. Later you'll pass them into their function calls:

```
pixels =224
BATCH_SIZE = 32
IMAGE_SIZE = (pixels, pixels)

datagen_kwargs = dict(rescale=1./255, validation_split=.20)
dataflow_kwargs = dict(target_size=IMAGE_SIZE,
batch_size=BATCH_SIZE,
interpolation="bilinear")
```

Create a generator object:

```
valid_datagen = tf.keras.preprocessing.image.ImageDataGenerator(
**datagen_kwargs)
```

Now you can specify the source directory from which this generator will stream the data. This generator will only stream 20% of the data, and this is designated as a validation dataset:

```
valid_generator = valid_datagen.flow_from_directory(
    data_dir, subset="validation", shuffle=False,
    **dataflow_kwargs)
```

You can use the same generator object for training data:

```
train_datagen = valid_datagen
train_generator = train_datagen.flow_from_directory(
data_dir, subset="training", shuffle=True, **dataflow_kwargs)
```

Inspect the output of the generator:

```
for image_batch, labels_batch in train_generator:
  print(image_batch.shape)
  print(labels_batch.shape)
  break

(32, 224, 224, 3)
(32, 5)
```

The output is represented as NumPy arrays. For a batch of images, the sample size is 32, with 224 pixels in height and width and three channels representing RGB color space. For the label batch, there are likewise 32 samples. Each row is one-hot encoded to represent which of the five classes it belongs to.

Another important thing to do is to retrieve the lookup dictionary of labels. During inferencing, the model will output the probability for each of the five classes. The only way to decode which class has the highest probability is with a prediction lookup dictionary of labels:

```
labels_idx = (train_generator.class_indices)
idx_labels = dict((v,k) for k,v in labels_idx.items())
print(idx_labels)

{0: 'daisy', 1: 'dandelion', 2: 'roses', 3: 'sunflowers',
4: 'tulips'}
```

A typical output from our classification model would be a NumPy array similar to this:

```
(0.7, 0.1, 0.1, 0.05, 0.05)
```

The position with the highest probability value is the first element. Map this index to the first key in idx_labels—in this case, daisy. This is how you capture the results of the prediction. Save the idx_labels dictionary:

```
import pickle
with open('prediction_lookup.pickle', 'wb') as handle:
    pickle.dump(idx_labels, handle,
    protocol=pickle.HIGHEST_PROTOCOL)
```

This is how to load it back:

```
with open('prediction_lookup.pickle', 'rb') as handle:
    lookup = pickle.load(handle)
```

Training the Model

Finally, for training you'll use a model built from a pretrained ResNet feature vector. This technique is known as *transfer learning*. TensorFlow Hub hosts many pretrained models for free. This is how to access it during your model construction process:

```
import tensorflow_hub as hub
NUM_CLASSES = 5
mdl = tf.keras.Sequential([
    tf.keras.layers.InputLayer(input_shape=IMAGE_SIZE + (3,)),
        hub.KerasLayer("https://tfhub.dev/google/imagenet/
resnet_v1_101/feature_vector/4", trainable=False),
tf.keras.layers.Dense(NUM_CLASSES, activation='softmax',
 name = 'custom_class')
])
mdl.build([None, 224, 224, 3])
```

The first layer is `InputLayer`. Remember that the expected input is $224 \times 224 \times 3$ pixels. You'll use the tuple addition trick to append an extra dimension to `IMAGE_SIZE`:

```
IMAGE_SIZE + (3,)
```

Now you have (224, 224, 3), which is a tuple that represents the dimension of an image as a NumPy array.

The next layer is the pretrained ResNet feature vector referenced by the URL to TensorFlow Hub. Let's use it as is so that we don't have to retrain it.

Next is the `Dense` layer with five nodes of output. Each output is the probability of the image belonging to that class. Then you'll build the model skeleton, with `None` as the first dimension. This means the first dimension, which represents the sample size of a batch, is not decided until runtime. This is how to handle batch input.

Inspect the model summary to make sure it's what you expected:

```
mdl.summary()
```

The output is shown in Figure 3-10.

```
Model: "sequential_1"
_________________________________________________________________
Layer (type)                 Output Shape              Param #
=================================================================
keras_layer_2 (KerasLayer)   (None, 2048)              42605504
_________________________________________________________________
custom_class (Dense)         (None, 5)                 10245
=================================================================
Total params: 42,615,749
Trainable params: 10,245
Non-trainable params: 42,605,504
_________________________________________________________________
```

Figure 3-10. Image classification model summary

Compile the model with `optimizers` and the corresponding `losses` function:

```
mdl.compile(
  optimizer=tf.keras.optimizers.SGD(lr=0.005, momentum=0.9),
  loss=tf.keras.losses.CategoricalCrossentropy(
from_logits=True,
label_smoothing=0.1),
  metrics=['accuracy'])
```

And then train it:

```
steps_per_epoch = train_generator.samples //
train_generator.batch_size
validation_steps = valid_generator.samples //
valid_generator.batch_size
mdl.fit(
    train_generator,
    epochs=5, steps_per_epoch=steps_per_epoch,
    validation_data=valid_generator,
    validation_steps=validation_steps)
```

You may see output similar to that in Figure 3-11.

```
Epoch 1/5
9/9 [==============================] - 22s 2s/step - loss: 1.5618 - accuracy: 0.3289 - val_loss: 1.4618 - val_accuracy: 0.4898
Epoch 2/5
9/9 [==============================] - 19s 2s/step - loss: 1.3874 - accuracy: 0.5639 - val_loss: 1.3818 - val_accuracy: 0.5425
Epoch 3/5
9/9 [==============================] - 19s 2s/step - loss: 1.2732 - accuracy: 0.6960 - val_loss: 1.2207 - val_accuracy: 0.7651
Epoch 4/5
9/9 [==============================] - 20s 2s/step - loss: 1.1901 - accuracy: 0.7981 - val_loss: 1.1935 - val_accuracy: 0.7772
Epoch 5/5
9/9 [==============================] - 20s 2s/step - loss: 1.1637 - accuracy: 0.8189 - val_loss: 1.1769 - val_accuracy: 0.7908
```

Figure 3-11. Output from training the image classification model

Summary

In this section, you learned how to process image files. Specifically, it is necessary to make sure you have a predetermined image size requirement set before you design the model. Once that standard is accepted, the next step is to resample images into that size and normalize the pixel's value into a smaller dynamic range. These routines are nearly universal. Also, streaming images into the training workflow is the most efficient method and a best practice, especially in cases where your working sample size approaches your Python runtime's memory.

Preparing Text Data for Processing

For text data, each word or character needs to be represented as a numerical integer. This process is known as *tokenization*. Further, if the goal is classification, then the target needs to be encoded as *classes*. If the goal is something more complicated, such as translation, then the target language in the training data (such as French in an English-to-French translation) also requires its own tokenization process. This is because the target is essentially a long string of text, just like the input text. Likewise, you also need to think about whether to tokenize the target at the word or character level.

Text data can be presented in many different formats. From a content organization perspective, it may be stored and organized as a table, with one column containing the body or string of text and another column containing labels, such as a binary sentiment indicator. It may be a free-form file, with lines of different lengths and a carriage return at the end of each line. It may be a manuscript in which blocks of text are defined by paragraphs or sections.

There are many ways to determine the processing techniques and logic to use as you set up a natural language processing (NLP) machine learning problem; this section will cover some of the most frequently used techniques.

This example will use text from William Shakespeare's tragedy *Coriolanus,* which is a simple public-domain example hosted on Google. You will build a *text generation model* that will learn how to write in Shakespeare's style.

Tokenizing Text

Text is represented by strings of characters. These characters need to be converted to integers for modeling tasks. This example is a raw text string from *Coriolanus.*

Let's import the necessary libraries and download the text file:

```
import tensorflow as tf
import numpy as np
import os
import time

FILE_URL = 'https://storage.googleapis.com/download.tensorflow.org/
data/shakespeare.txt'
FILE_NAME = 'shakespeare.txt'
path_to_file = tf.keras.utils.get_file('shakespeare.txt', FILE_URL)
```

Open it and output a few lines of sample text:

```
text = open(path_to_file, 'rb').read().decode(encoding='utf-8')
print ('Length of text: {} characters'.format(len(text)))
```

Inspect this text by printing the first 400 characters:

```
print(text[:400])
```

The output is shown in Figure 3-12.

To tokenize each character in this file, a simple `set` operation will suffice. This operation will create a unique set of characters found in the text string:

```
vocabulary = sorted(set(text))
print ('There are {} unique characters'.format(len(vocabulary)))
```

```
There are 65 unique characters
```

A glimpse of the list `vocabulary` is shown in Figure 3-13.

```
First Citizen:
Before we proceed any further, hear me speak.

All:
Speak, speak.

First Citizen:
You are all resolved rather to die than to famish?

All:
Resolved. resolved.

First Citizen:
First, you know Caius Marcius is chief enemy to the people.

All:
We know't, we know't.

First Citizen:
Let us kill him, and we'll have corn at our own price.
Is't a verdict?

All:
No more talking on't; let it
```

Figure 3-12. Sample of William Shakespeare's Coriolanus

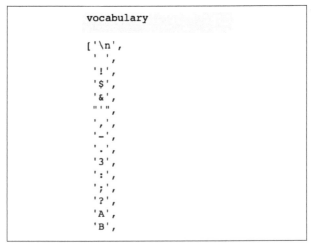

Figure 3-13. Part of the vocabulary list from Coriolanus

These tokens include punctuation, as well as both upper- and lowercase characters. It is not always necessary to include both upper- and lowercase characters; if you don't want to, you can convert every character to lowercase before performing the set operation. Since you sorted the token list, you can see that special characters are also being tokenized. In some cases, this is not necessary; these tokens can be removed manually. The following code will convert all characters to lowercase and then perform the set operation:

```
vocabulary = sorted(set(text.lower()))
print ('There are {} unique characters'.format(len(vocabulary)))
```

```
There are 39 unique characters
```

You might be wondering if it would be reasonable to tokenize text at the word level instead of the character level. After all, the word is the fundamental unit of semantic understanding of a text string. Although this reasoning is sound and has some logic to it, in reality it creates more work and problems, while not really adding value to the training process or accuracy to the model. To illustrate why, let's try to tokenize the text string by words. The first thing to recognize is that words are separated by spaces. So you need to split the text string on spaces:

```
vocabulary_word = sorted(set(text.lower().split(' ')))
print ('There are {} unique words'.format(len(vocabulary_word)))
```

```
There are 41623 unique words
```

Inspect the list vocabulary_word, shown in Figure 3-14.

With special characters and carriage returns embedded in each word token, this list is nearly unusable. It would require considerable work to clean it up with regular expressions or more sophisticated logic. In some cases, punctuation marks are attached to words. Further, the list of word tokens is much larger than the character-level token list. This makes it much more complicated for the model to learn the patterns in the text. For these reasons and the lack of proven benefit, it is not a common practice to tokenize text at the word level. If you wish to use word-level tokenization, then it is common to perform a word-embedding operation to reduce the variability and dimensions of the utterance representations.

```
['',
 '\ntwice',
 '\nwas',
 '&c.\n\nmontague:\nand',
 "&c:\nwe'll",
 '&c:\nweapons,',
 "'",
 "'?\n\nthird",
 "'a",
 "'aged",
 "'alas,",
 "'alas,'\ni",
 "'all",
 "'almost",
 "'ay",
 "'ay'",
 "'ay,'",
 "'ay,'\nand",
```

Figure 3-14. Sample of tokenized words

Creating a Dictionary and Reverse Dictionary

Once you have the list of tokens that contains the chosen characters, you'll need to map each token to an integer. This is
known as the *dictionary*. Likewise, you'll need to create a
reverse dictionary that maps the integers back to the tokens.

Generating an integer is easy with the enumerate function. This
function takes a list as input and returns an integer corresponding to each unique element in the list. In this case, the list
contains tokens:

```
for i, u in enumerate(vocabulary):
  print(i, u)
```

You can see a sample of this result in Figure 3-15.

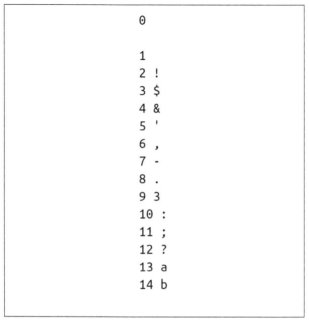

Figure 3-15. Sample enumerated output of a token list

Next you need to make this into a dictionary. A dictionary is really a collection of key-value pairs used as a lookup table: when you give it a key, it returns the value corresponding to that key. The notation to build a dictionary, with the key being the token and the value being the integer, is:

```
char_to_index = {u:i for i, u in enumerate(vocabulary)}
```

The output will look like Figure 3-16.

This dictionary is used to convert text into integers. At inference time, the model output is also in the format of integers. Therefore, if you want the output as text, then you'll need a reverse dictionary to map the integers back to characters. To do this, simply reverse the order of i and u:

```
index_to_char = {i:u for i, u in enumerate(vocabulary)}
```

```
{'\n': 0,
  ' ': 1,
  '!': 2,
  '$': 3,
  '&': 4,
  '"': 5,
  ',': 6,
  '_': 7,
  '.': 8,
  '3': 9,
  ':': 10,
  ';': 11,
  '?': 12,
  'a': 13,
  'b': 14,
```

Figure 3-16. Sample of character-to-index dictionary

Tokenization is the most basic and necessary step in most NLP problems. A text generation model will not generate plain text as the output; it generates the output in a series of integers. In order for this series of indices to map to letters (tokens), you need a lookup table. index_to_char is specifically built for this purpose. Using index_to_char, you can look up each character (token) by key, where the key is the index from the model's output. Without index_to_char, you will not be able to map model outputs back to a readable, plain-text format.

Wrapping Up

In this chapter, you learned how to handle some of the most common data structures: tables, images, and text. Tabular datasets (the structured, CSV-style data) are very common, are returned from a typical database query, and are frequently used as training data. You learned how to deal with columns of different data types in such structures, as well as how to model feature interactions by crossing columns of interest.

For image data, you learned that you need to standardize image size and pixel values before using the image collection as a whole to train a model, and that you need to keep track of image labels.

Text data is by far the most diverse data type, in terms of both format and use. Nevertheless, whether the data is for text classification, translation, or question-and-answer models, tokenization and dictionary construction processes are very common. The methods and approaches described in this chapter are by no means exhaustive or comprehensive; rather, they represent "table stakes" when dealing with these data types.

## CHAPTER 4

Reusable Model Elements

Developing an ML model can be a daunting task. Besides the data engineering aspect of the task, you also need to understand how to build the model. In the early days of ML, tree-based models (such as random forests) were king for applying straight-up classification or regression tasks to tabular datasets, and model architecture was determined by parameters related to model initialization. These parameters, known as hyperparameters, include the number of decision trees in a forest and the number of features considered by each tree when splitting a node. However, it is not straightforward to convert some types of data, such as images or text, into tabular form: images may have different dimensions, and texts vary in length. That's why deep learning has become the de facto standard model architecture for image and text classification.

As deep-learning architecture gains popularity, a community has grown around it. Creators have built and tested different model structures for academic and Kaggle challenges. Many have made their models open source so that they are available for transfer learning—anyone can use them for their own purposes.

For example, ResNet is an image classification model trained on the ImageNet dataset, which is about 150GB in size and contains more than a million images. The labels in this data include plants, geological formations, natural objects, sports, persons, and animals. So how can you reuse the ResNet model to classify your own set of images, even with different categories or labels?

Open source models such as ResNet have very complicated structures. While the source code is available for anyone to access on sites like GitHub, downloading the source code is not the most user-friendly way to reproduce or reuse these models. There are almost always other dependencies that you have to overcome to compile or run the source code. So how can we make such models available and usable to nonexperts?

TensorFlow Hub (TFH) is designed to solve this problem. It enables transfer learning by making a variety of ML models freely available as libraries or web API calls. Anyone can write just a single line of code to load the model. All models can be invoked via a simple web call, and then the entire model is downloaded to your source code's runtime. You don't need to build the model yourself.

This definitely saves development and training time and increases accessibility. It also allows users to try out different models and build their own applications more quickly. Another benefit of transfer learning is that since you are not retraining the whole model from scratch, you may not need a high-powered GPU or TPU to get started.

In this chapter, we are going to take a look at just how easy it is to leverage TensorFlow Hub. So let's start with how TFH is organized. Then you'll download one of the TFH pretrained image classification models and see how to use it for your own images.

The Basic TensorFlow Hub Workflow

TensorFlow Hub (*https://oreil.ly/dQxxy*) (Figure 4-1) is a repository of pretrained models curated by Google. Users may download any model into their own runtime and perform fine-tuning and training with their own data.

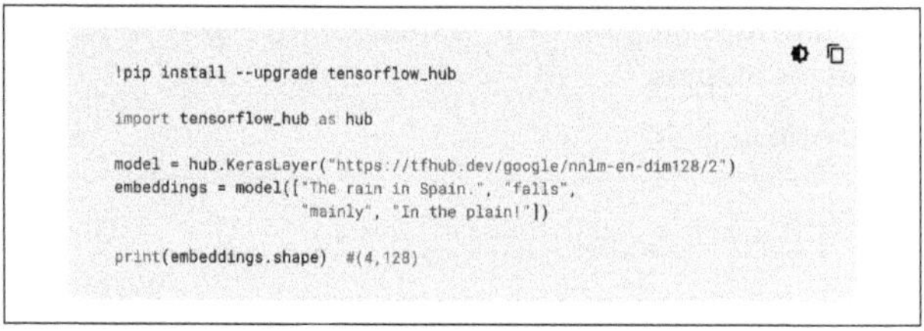

Figure 4-1. TensorFlow Hub home page

To use TFH, you must install it via the familiar Pythonic `pip install` command in your Python cell or terminal:

```
pip install --upgrade tensorflow_hub
```

Then you can start using it in your source code by importing it:

```
import tensorflow_hub as hub
```

First, invoke the model:

```
model = hub.KerasLayer("https://tfhub.dev/google/nnlm-en-dim128/2")
```

This is a pretrained text embedding model. *Text embedding* is the process of mapping a string of text to a multidimensional vector of numeric representation. You can give this model four text strings:

```
embeddings = model(["The rain in Spain.", "falls",
                    "mainly", "In the plain!"])
```

Before you look at the results, inspect the shape of the model output:

```
print(embeddings.shape)
```

It should be:

```
(4, 128)
```

There are four outputs, each 128 units long. Figure 4-2 shows one of the outputs:

```
print(embeddings[0])
```

```
tf.Tensor(
[ 1.59735829e-01 -1.29867658e-01 -1.35819957e-01 -9.24930871e-02
  2.10374016e-02 -7.29679167e-02 -1.11245513e-01 -9.90176424e-02
  .....
  5.64674214e-02  8.22738558e-03 -6.20092526e-02  7.22365454e-03
  9.19964239e-02  3.05843651e-02  1.27910644e-01  3.83963063e-02], shape=(128,), dtype=float32)
```

Figure 4-2. Text embedding output

As indicated in this simple example, you did not train this model. You only loaded it and used it to get a result with your own data. This pretrained model simply converts each text string into a vector representation of 128 dimensions.

On the TensorFlow Hub home page, click the Models tab. As you can see, TensorFlow Hub categorizes its pretrained models into four problem domains: image, text, video, and audio.

Figure 4-3 shows the general pattern for a transfer learning model.

From Figure 4-3, you can see that the pretrained model (from TensorFlow Hub) is sandwiched between an input layer and an output layer, and there can be some optional layers prior to the output layer.

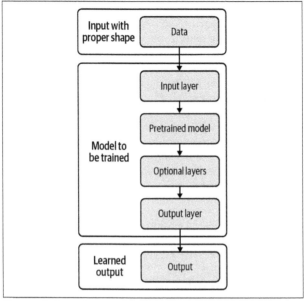

Figure 4-3. General pattern for transfer learning

To use any of the models, you'll need to address a few important considerations, such as input and output:

Input layer

Input data must be properly formatted (or "shaped"), so pay special attention to each model's input requirements (found in the Usage section on the web page that describes the individual model). Take the ResNet feature vector (*https://oreil.ly/6xGeP*), for example: the Usage section states the required size and color values for the input images and that the output is a batch of feature vectors. If your data does not meet the requirements, you'll need to apply some of the data transformation techniques you learned in "Preparing Image Data for Processing" on page 47.

Output layer

Another important and necessary element is the output layer. This is a must if you wish to retrain the model with your own data. In the simple embedding example shown earlier, we didn't retrain the model; we merely fed it a few text strings to see the model output. An output layer serves the purpose of mapping the model output to the most likely labels if the problem is a classification problem. If it is a regression problem, then it serves to map the model output to a numeric value. A typical output layer is called "dense," with either one node (for regression or binary classification) or multiple nodes (such as for multiclass classification).

Optional layers

Optionally, you can add one or more layers before the output layer to improve model performance. These layers may help you extract more features to improve model accuracy, such as a convolution layer (Conv1D, Conv2D). They can also help prevent or reduce model overfitting. For example, dropout reduces overfitting by randomly setting an output to zero. If a node outputs an array such as [0.5, 0.1, 2.1, 0.9] and you set a dropout ratio of 0.25, then during training, by random chance, one of the four values in the array will be set to zero; for example, [0.5, 0, 2.1, 0.9]. Again, this is considered optional. Your training does not require it, but it may help improve your model's accuracy.

Image Classification by Transfer Learning

We are going to walk through an image classification example with transfer learning. In this example, your image data consists of five classes of flowers. You will use the ResNet feature vector as your pretrained model. We will address these common tasks:

- Model requirements
- Data transformation and input processing
- Model implementation with TFH
- Output definition
- Mapping output to plain-text format

Model Requirements

Let's look at the ResNet v1_101 feature vector (*https://oreil.ly/70grM*) model. This web page contains an overview, a download URL, instructions, and, most importantly, the code you'll need to use the model.

In the Usage section, you can see that to load the model, all you need to do is pass the URL to hub.KerasLayer. The Usage section also includes the model requirements. By default, it expects the input image, which is written as an array of shape [height, width, depth], to be [224, 224, 3]. The pixel value is expected to be within the range [0, 1]. As the output, it provides the Dense layer with the number of nodes, which reflects the number of classes in the training images.

Data Transformation and Input Processing

It is your job to transform your images into the required shape and normalize the pixel scale to within the required range. As we've seen, images usually come in different size and pixel values. A typical color JPEG image pixel value for each RGB channel might be anywhere from 0 to 225. So, we need operations to standardize image size to [224, 224, 3], and to normalize pixel value to a [0, 1] range. If we use ImageDataGenerator in TensorFlow, these operations are provided as input flags. Here's how to load the images and create a generator:

1. Start by loading the libraries:

```
import tensorflow as tf
import tensorflow_hub as hub
import numpy as np
import matplotlib.pylab as plt
```

2. Load the data you need. For this example, let's use the flower images provided by TensorFlow:

```
data_dir = tf.keras.utils.get_file(
    'flower_photos',
'https://storage.googleapis.com/download.tensorflow.org/
example_images/flower_photos.tgz',
    untar=True)
```

3. Open `data_dir` and find the images. You can see the file structure in the file path:

```
!ls -lrt /root/.keras/datasets/flower_photos
```

Here is what will display:

```
total 620
-rw-r----- 1 270850 5000 418049 Feb  9  2016 LICENSE.txt
drwx------ 2 270850 5000  45056 Feb 10  2016 tulips
drwx------ 2 270850 5000  40960 Feb 10  2016 sunflowers
drwx------ 2 270850 5000  36864 Feb 10  2016 roses
drwx------ 2 270850 5000  53248 Feb 10  2016 dandelion
drwx------ 2 270850 5000  36864 Feb 10  2016 daisy
```

There are five classes of flowers. Each class corresponds to a directory.

4. Define some global variables to store pixel values and *batch size* (the number of samples in a batch of training images). You don't yet need the third dimension of the image, just the height and width for now:

```
pixels =224
BATCH_SIZE = 32
IMAGE_SIZE = (pixels, pixels)
NUM_CLASSES = 5
```

5. Specify image normalization and a fraction of data for cross validation. It is a good idea to hold out a fraction of training data for cross validation, which is a means of evaluating the model training process through each epoch. At the end of each training epoch, the model contains a set of trained weights and biases. At this point, the data held out for cross validation, which the model has never seen, can be used as a test for model accuracy:

```
datagen_kwargs = dict(rescale=1./255, validation_split=.20)
dataflow_kwargs = dict(target_size=IMAGE_SIZE,
batch_size=BATCH_SIZE,
interpolation="bilinear")

valid_datagen = tf.keras.preprocessing.image.
ImageDataGenerator(
    **datagen_kwargs)
valid_generator = valid_datagen.flow_from_directory(
    data_dir, subset="validation", shuffle=False,
    **dataflow_kwargs)
```

The ImageDataGenerator definition and generator instance both accept our arguments in a dictionary format. The rescaling factor and validation fraction go to the generator definition, while the standardized image size and batch size go to the generator instance.

The interpolation argument indicates that the generator needs to resample the image data to target_size, which is 224 × 224 pixels.

Now, do the same for the training data generator:

```
train_datagen = valid_datagen
train_generator = train_datagen.flow_from_directory(
    data_dir, subset="training", shuffle=True,
    **dataflow_kwargs)
```

6. Identify mapping of class index to class name. Since the flower classes are encoded in the index, you need a map to recover the flower class names:

```
labels_idx = (train_generator.class_indices)
idx_labels = dict((v,k) for k,v in labels_idx.items())
```

You can display the `idx_labels` to see how these classes are mapped:

```
idx_labels

{0: 'daisy', 1: 'dandelion', 2: 'roses', 3: 'sunflowers',
4: 'tulips'}
```

You've now normalized and standardized your image data. The image generators are defined and instantiated for training and validation data. You also have the label lookup to decode model prediction, and you're ready to implement the model with TFH.

Model Implementation with TensorFlow Hub

As you saw back in Figure 4-3, the pretrained model is sandwiched between an input and an output layer. You can define this model structure accordingly:

```
model = tf.keras.Sequential([
    tf.keras.layers.InputLayer(input_shape=IMAGE_SIZE + (3,)),
hub.KerasLayer("https://tfhub.dev/google/imagenet/resnet_v1_101/
feature_vector/4", trainable=False),
    tf.keras.layers.Dense(NUM_CLASSES, activation='softmax',
name = 'flower_class')
])

model.build([None, 224, 224, 3]) !!C04!!
```

Notice a few things here:

- There is an input layer that defines the input shape of images as [224, 224, 3].

- When `InputLayer` is invoked, `trainable` should be set to False. This indicates that you want to reuse the current values from the pretrained model.

- There is an output layer called `Dense` that provides the model output (this is described in the Usage section of the summary page).

After the model is built, you're ready to start training. First, specify the loss function and pick an optimizer:

```
model.compile(
  optimizer=tf.keras.optimizers.SGD(lr=0.005, momentum=0.9),
loss=tf.keras.losses.CategoricalCrossentropy(
from_logits=True,
label_smoothing=0.1),
metrics=['accuracy'])
```

Then specify the number of batches for training data and cross-validation data:

```
steps_per_epoch = train_generator.samples //
train_generator.batch_size
validation_steps = valid_generator.samples //
valid_generator.batch_size
```

Then start the training process:

```
model.fit(
    train_generator,
    epochs=5, steps_per_epoch=steps_per_epoch,
    validation_data=valid_generator,
    validation_steps=validation_steps)
```

After the training process runs through all the epochs specified, the model is trained.

Defining the Output

According to the Usage guideline, the output layer `Dense` consists of a number of nodes, which reflects how many classes are in the expected images. This means each node outputs a probability for that class. It is your job to find which one of these probabilities is the highest and map that node to the flower class using `idx_labels`. Recall that the `idx_labels` dictionary looks like this:

```
{0: 'daisy', 1: 'dandelion', 2: 'roses', 3: 'sunflowers',
4: 'tulips'}
```

The `Dense` layer's output consists of five nodes in the exact same order. You'll need to write a few lines of code to map the position with the highest probability to the corresponding flower class.

Mapping Output to Plain-Text Format

Let's use the validation images to understand a bit more about
how to map the model prediction output to the actual class for
each image. You'll use the `predict` function to score these validation images. Retrieve the NumPy array for the first batch:

```
sample_test_images, ground_truth_labels = next(valid_generator)

prediction = model.predict(sample_test_images)
```

There are 731 images and 5 corresponding classes in the cross-
validation data. Therefore, the output shape is [731, 5]:

```
array([[9.9994004e-01, 9.4704428e-06, 3.8405190e-10, 5.0486942e-05,
        1.0701914e-08],
       [5.9500107e-06, 3.1842374e-06, 3.5622744e-08, 9.9999082e-01,
        3.0683900e-08],
       [9.9994218e-01, 5.9974178e-07, 5.8693445e-10, 5.7049790e-05,
        9.6709634e-08],
       ...,
       [3.1268091e-06, 9.9986601e-01, 1.5343730e-06, 1.2935932e-04,
        2.7383029e-09],
       [4.8439368e-05, 1.9247003e-05, 1.8034354e-01, 1.6394027e-02,
        8.0319476e-01],
       [4.9799957e-07, 9.9232978e-01, 3.5823192e-08, 7.6697678e-03,
        1.7666844e-09]], dtype=float32)
```

Each row represents the probability distribution for the image
class. For the first image, the highest probability, 1.0701914e-08
(highlighted in the preceding code), is in the last position,
which corresponds to index 4 of that row (remember, the numbering of an index starts with 0).

Now you need to find the position where the highest probability occurs for each row, using this code:

```
predicted_idx = tf.math.argmax(prediction, axis = -1)
```

And if you display the results with the `print` command, you'll see this:

```
print (predicted_idx)

<tf.Tensor: shape=(731,), dtype=int64, numpy=
array([0, 3, 0, 1, 0, 4, 4, 1, 2, 3, 4, 1, 4, 0, 4, 3, 1, 4, 4, 0,
       ...
       3, 2, 1, 4, 1])>
```

Now, apply the lookup with `idx_labels` to each element in this array. For each element, use a function:

```
def find_label(idx):
    return idx_labels[idx]
```

To apply a function to each element of a NumPy array, you'll need to vectorize the function:

```
find_label_batch = np.vectorize(find_label)
```

Then apply this vectorized function to each element in the array:

```
result = find_label_batch(predicted_idx)
```

Finally, output the result side by side with the image folder and filename so that you can save it for reporting or further investigation. You can do this with Python pandas DataFrame manipulation:

```
import pandas as pd
predicted_label = result_class.tolist()
file_name = valid_generator.filenames

results=pd.DataFrame({"File":file_name,
                      "Prediction":predicted_label})
```

Let's take a look at the `results` dataframe, which is 731 rows × 2 columns.

	File	Prediction
0	daisy/100080576_f52e8ee070_n.jpg	daisy
1	daisy/10140303196_b88d3d6cec.jpg	sunflowers
2	daisy/10172379554_b296050f82_n.jpg	daisy
3	daisy/10172567486_2748826a8b.jpg	dandelion
4	daisy/10172636503_21bededa75_n.jpg	daisy
...	...	...
726	tulips/14068200854_5c13668df9_m.jpg	sunflowers
727	tulips/14068295074_cd8b85bffa.jpg	roses
728	tulips/14068348874_7b36c99f6a.jpg	dandelion
729	tulips/14068378204_7b26baa30d_n.jpg	tulips
730	tulips/14071516088_b526946e17_n.jpg	dandelion

Evaluation: Creating a Confusion Matrix

A *confusion matrix*, which evaluates the classification results by comparing model output with ground truth, is the easiest way to get an initial feel for how well the model performs. Let's look at how to create a confusion matrix.

You'll use pandas Series as the data structure for building your confusion matrix:

```
y_actual = pd.Series(valid_generator.classes)
y_predicted = pd.Series(predicted_idx)
```

Then you'll utilize pandas again to produce the matrix:

```
pd.crosstab(y_actual, y_predicted, rownames = ['Actual'],
colnames=['Predicted'], margins=True)
```

Figure 4-4 shows the confusion matrix. Each row represents the distribution of actual flower labels by predictions. For example, looking at the first row, you will notice that there are a total of 126 samples that are actually class 0, which is daisy. The model correctly predicted 118 of these images as class 0; four are misclassified as class 1, which is dandelion; one is misclassified as class 2, which is roses; three are misclassified as class 3, which is sunflowers; and none has been misclassified as class 4, which is tulips.

Predicted	0	1	2	3	4	All
Actual						
0	118	4	1	3	0	126
1	7	156	4	8	4	179
2	1	0	110	2	15	128
3	2	6	6	122	3	139
4	3	1	9	9	137	159
All	131	167	130	144	159	731

Figure 4-4. Confusion matrix for flower image classification

Next, use the `sklearn` library to provide a statistical report for each class of images:

```
from sklearn.metrics import classification_report
report = classification_report(truth, predicted_results)
print(report)
              precision    recall  f1-score   support

           0       0.90      0.94      0.92       126
           1       0.93      0.87      0.90       179
           2       0.85      0.86      0.85       128
           3       0.85      0.88      0.86       139
           4       0.86      0.86      0.86       159

    accuracy                           0.88       731
   macro avg       0.88      0.88      0.88       731
weighted avg       0.88      0.88      0.88       731
```

This result shows that the model has the best performance when classifying daisies (class 0), with an f1-score of 0.92. Its performance is worst in classifying roses (class 2), with an f1-score of 0.85. The "support" column indicates the sample size in each class.

Summary

You have just completed an example project using a pretrained model from TensorFlow Hub. You appended the necessary input layer, performed data normalization and standardization, trained the model, and scored a batch of images.

This experience shows the importance of meeting the model's input and output requirements. Just as importantly, pay close attention to the output format of the pretrained model. (This information is all available in the model documentation page on the TensorFlow Hub website.) Finally, you also need to create a function that maps the model output to plain text to make it meaningful and interpretable.

Using the tf.keras.applications Module for Pretrained Models

Another place to find a pretrained model for your own use is the tf.keras.applications module (see the list of available models (*https://oreil.ly/HQJBl*)). When the Keras API became available in TensorFlow, this module became a part of the TensorFlow ecosystem.

Each model comes with pretrained weights, and using them is just as easy as using TensorFlow Hub. Keras provides the flexibility needed to conveniently fine-tune your models. By making each layer in a model accessible, tf.keras.applications lets you specify which layers to retrain and which layers to leave untouched.

Model Implementation with tf.keras.applications

As with TensorFlow Hub, you need only one line of code to load a pretrained model from the Keras module:

```
base_model = tf.keras.applications.ResNet101V2(
input_shape = (224, 224, 3),
include_top = False,
weights = 'imagenet')
```

Notice the include_top input argument. Remember that you need to add an output layer for your own data. By setting include_top to False, you can add your own Dense layer for the classification output. You'll also initialize the model weights from imagenet.

Then place base_model inside a sequential architecture, as you did in the TensorFlow Hub example:

```
model2 = tf.keras.Sequential([
  base_model,
  tf.keras.layers.GlobalAveragePooling2D(),
  tf.keras.layers.Dense(NUM_CLASSES,
  activation = 'softmax',
  name = 'flower_class')
])
```

Add GlobalAveragePooling2D, which averages the output array into one numeric value, to do an aggregation before sending it to the final Dense layer for prediction.

Now compile the model and launch the training process as usual:

```
model2.compile(
  optimizer=tf.keras.optimizers.SGD(lr=0.005, momentum=0.9),
  loss=tf.keras.losses.CategoricalCrossentropy(
  from_logits=True, label_smoothing=0.1),
  metrics=['accuracy']
)

model2.fit(
    train_generator,
    epochs=5, steps_per_epoch=steps_per_epoch,
    validation_data=valid_generator,
    validation_steps=validation_steps)
```

To score image data, follow the same steps as you did in "Mapping Output to Plain-Text Format" on page 76.

Fine-Tuning Models from tf.keras.applications

If you wish to experiment with your training routine by releasing some layers of the base model for training, you can do so easily. To start, you need to find out exactly how many layers are in your base model and designate the base model as trainable:

```
print("Number of layers in the base model: ",
      len(base_model.layers))
base_model.trainable = True

Number of layers in the base model:  377
```

As indicated, in this version of the ResNet model, there are 377 layers. Usually we start the retraining process with layers close to the end of the model. In this case, designate layer 370 as the starting layer for fine-tuning, while holding the weights in layers before 300 untouched:

```
fine_tune_at = 370

for layer in base_model.layers[: fine_tune_at]:
  layer.trainable = False
```

Then put together the model with the Sequential class:

```
model3 = tf.keras.Sequential([
  base_model,
  tf.keras.layers.GlobalAveragePooling2D(),
  tf.keras.layers.Dense(NUM_CLASSES,
  activation = 'softmax',
  name = 'flower_class')
])
```

TIP

You can try tf.keras.layers.Flatten() instead of tf.keras.layers.GlobalAveragePooling2D(), and see which one gives you a better model.

Compile the model, designating the optimizer and loss function as you did with TensorFlow Hub:

```
model3.compile(
  optimizer=tf.keras.optimizers.SGD(lr=0.005, momentum=0.9),
  loss=tf.keras.losses.CategoricalCrossentropy(
  from_logits=True,
  label_smoothing=0.1),
  metrics=['accuracy']
)
```

Launch the training process:

```
fine_tune_epochs = 5
steps_per_epoch = train_generator.samples //
train_generator.batch_size
validation_steps = valid_generator.samples //
valid_generator.batch_size
model3.fit(
    train_generator,
    epochs=fine_tune_epochs,
    steps_per_epoch=steps_per_epoch,
    validation_data=valid_generator,
    validation_steps=validation_steps)
```

This training may take considerably longer, since you've freed up more layers from the base model for retraining. Once the training is done, score the test data and compare the results as described in "Mapping Output to Plain-Text Format" on page 76 and "Evaluation: Creating a Confusion Matrix" on page 78.

Wrapping Up

In this chapter, you learned how to conduct transfer learning using pretrained, deep-learning models. There are two convenient ways to access pretrained models: TensorFlow Hub and the tf.keras.applications module. Both are simple to use and have elegant APIs and styles for quick model development. However, users are responsible for shaping their input data correctly and for providing a final Dense layer to handle model output.

There are plenty of freely accessible pretrained models with abundant inventories that you can use to work with your own data. Taking advantage of them using transfer learning lets you spend less time building, training, and debugging models.

## CHAPTER 5

Data Pipelines for Streaming Ingestion

Data ingestion is an important part of your workflow. There are several steps to perform before raw data is in the correct input format expected by the model. These steps are known as the *data pipeline*. Steps in a data pipeline are important because they will also be applied to the production data, which is the data consumed by the model when the model is deployed. Whether you are in the process of building and debugging a model or getting it ready for deployment, you need to format the raw data for the model's consumption.

It is important to use the same series of steps in the model-building process as you do in deployment planning, so that the test data is processed the same way as the training data.

In Chapter 3 you learned how the Python generator works, and in Chapter 4 you learned how to use the `flow_from_directory` method for transfer learning. In this chapter, you will see more of the tools that TensorFlow provides to handle other data types, such as text and numeric arrays. You'll also learn how to handle another type of file structure for images. File organization becomes especially important when handling text or images for model training because it is common to use directory names as labels. This chapter will recommend a practice for directory organization when it comes to building and training a text or image classification model.

Streaming Text Files with the text_dataset_from_directory Function

You can stream pretty much any files in a pipeline, as long as you organize the directory structure correctly. In this section we'll look at an example using text files, which would apply in use cases such as text classification and sentiment analysis. Here we are interested in the `text_dataset_from_directory` function, which works similarly to the `flow_from_directory` method that we used for streaming images.

In order to use this function for a text classification problem, you have to follow the directory organization described in this section. In your current working directory, you must have subdirectories with names that match the labels or class names for your text. For example, if you are doing text classification model training, you have to organize your training texts into positives and negatives. This is the process of training data labeling; it has to be done to set up the data for the model to learn what a positive or negative comment looks like. If the text is a corpus of movie reviews classified as positive or negative, then the subdirectory names might be *pos* and *neg*. Within each subdirectory, you have all the text files for that class. Therefore, your directory structure would be similar to this:

```
Current working directory
    pos
        p1.txt
        p2.txt
    neg
        n1.txt
        n2.txt
```

As an example, let's try building a data ingestion pipeline for text data using a corpus of movie reviews from the Internet Movie Database (IMDB).

Downloading Text Data and Setting Up Directories

The text data you will use for this section is the Large Movie Review Dataset (*https://oreil.ly/EabEP*). You can download it directly or use the get_file function to do so. Let's start by importing the necessary libraries and then downloading the file:

```
import io
import os
import re
import shutil
import string
import tensorflow as tf

url = "https://ai.stanford.edu/~amaas/data/sentiment/
        aclImdb_v1.tar.gz"

ds = tf.keras.utils.get_file("aclImdb_v1.tar.gz", url,
                                untar=True, cache_dir='.',
                                cache_subdir='')
```

Conveniently, by passing untar=True, the get_file function also decompresses the file. This will create a directory called *aclImdb* in the current directory. Let's encode this file path as a variable for future reference:

```
ds_dir = os.path.join(os.path.dirname(ds), 'aclImdb')
```

List this directory to see what's inside:

```
train_dir = os.path.join(ds_dir, 'train')
os.listdir(train_dir)

['neg',
 'unsup',
 'urls_neg.txt',
 'urls_unsup.txt',
 'pos',
 'urls_pos.txt',
 'unsupBow.feat',
 'labeledBow.feat']
```

There is one directory (*unsup*) not in use, so you'll need to get rid of it:

```
unused_dir = os.path.join(train_dir, 'unsup')
shutil.rmtree(unused_dir)
```

Now take a look at the content in the training directory:

```
!ls -lrt ./aclImdb/train
-rw-r--r-- 1 7297 1000  2450000 Apr 12  2011 urls_unsup.txt
drwxr-xr-x 2 7297 1000   364544 Apr 12  2011 pos
drwxr-xr-x 2 7297 1000   356352 Apr 12  2011 neg
-rw-r--r-- 1 7297 1000   612500 Apr 12  2011 urls_pos.txt
-rw-r--r-- 1 7297 1000   612500 Apr 12  2011 urls_neg.txt
-rw-r--r-- 1 7297 1000 21021197 Apr 12  2011 labeledBow.feat
-rw-r--r-- 1 7297 1000 41348699 Apr 12  2011 unsupBow.feat
```

The two directories are *pos* and *neg*. These names will be encoded as categorical variables in the text classification task.

It's important to clean up your subdirectories and ensure that all directories contain text for the classification training. If we had not removed that unused directory, its name would have become a categorical variable, which is not our intention at all. The other files there are fine and don't impact the outcome here. Again, remember that directory names are used as labels, so make sure you have *only* directories that are intended for the model to learn and map to labels.

Creating the Data Pipeline

Now that your files are properly organized, you're ready to create the data pipeline. Let's set up a few variables:

```
batch_size = 1024
seed = 123
```

The batch size tells the generator how many samples to use in one iteration of training. It's also a good idea to assign a seed so that each time you execute the generator, it streams the files in the same order. Without the seed assignment, the generator will output the files in random order.

Then define a pipeline using the `test_dataset_from_directory` function. It will return a dataset object:

```
train_ds = tf.keras.preprocessing.text_dataset_from_directory(
    'aclImdb/train', batch_size=batch_size, validation_split=0.2,
    subset='training', seed=seed)
```

In this case, the directory that contains subdirectories is *aclImdb/train*. This pipeline definition is for 80% of the training dataset, which is designated by `subset='training'`. The other 20% is reserved for cross validation.

For the cross-validation data, you'll define the pipeline in a similar fashion:

```
val_ds = tf.keras.preprocessing.text_dataset_from_directory(
    'aclImdb/train', batch_size=batch_size, validation_split=0.2,
    subset='validation', seed=seed)
```

Once you execute the two pipelines in the preeding code, this is the expected output:

```
Found 25000 files belonging to 2 classes.
Using 20000 files for training.
Found 25000 files belonging to 2 classes.
Using 5000 files for validation.
```

Because there are two subdirectories in *aclImdb/train*, the generator recognizes them as classes. And because of the 20% split, 5,000 files are held for cross validation.

Inspecting the Dataset

With the generator in place, let's take a look at the contents of these files. The way to inspect a TensorFlow dataset is to iterate through it and select a few samples. The following code snippet takes the first batch of samples and then randomly selects five rows of movie reviews:

```
import random
idx = random.sample(range(1, batch_size), 5)
for text_batch, label_batch in train_ds.take(1):
  for i in idx:
    print(label_batch[i].numpy(), text_batch.numpy()[i])
```

Here, `idx` is a list that holds five randomly generated integers within the range of `batch_size`. Then `idx` is used as the index to select the text and label from the dataset.

The dataset will yield a tuple consisting of `text_batch` and `label_batch`; a tuple is useful here because it keeps track of the text and its label (class). These are five randomly selected rows of text and corresponding labels:

```
1 b'Very Slight Spoiler<br /><br /> This movie (despite being….
1 b"Not to mention easily Pierce Brosnan's best performance….
0 b'Bah. Another tired, desultory reworking of an out of copyright…
0 b'All the funny things happening in this sitcom is based on the…
0 b'This is another North East Florida production, filmed mainly…
```

The first two are positive reviews (indicated by the digit 1), and the last three are negative reviews (indicated by 0). This method is called *grouping by class*.

Summary

In this section, you learned how to stream text datasets. The method is similar to how images are streamed, with the exception of using the `text_dataset_from_directory` function. You learned grouping by class and the recommended directory organization for your data, which is important because directory names are used as labels for the model training process. In both image and text classification, you saw that directory names are used as labels.

Streaming Images with a File List Using the flow_from_dataframe Method

How your data is organized affects how you deal with the data ingestion pipeline. This is especially important with image data. During the image classification task in Chapter 4, you saw how different types of flowers were organized into directories with names corresponding to each flower type.

Grouping by class is not the only file organization method you will encounter in the real world. In another common style, shown in Figure 5-1, all images are thrown into one directory (which means it doesn't matter what you name the directory).

Figure 5-1. Another directory structure for storing image files

In this organization, you see that at the same level as the directory *flowers*, which contains all the images, there is a CSV file called *all_labels.csv*. This file contains two columns: one with all the filenames and one with the labels for those files:

```
file_name,label
7176723954_e41618edc1_n.jpg,sunflowers
2788276815_8f730bd942.jpg,roses
6103898045_e066cdeedf_n.jpg,dandelion
1441939151_b271408c8d_n.jpg,daisy
2491600761_7e9d6776e8_m.jpg,roses
```

To use image files stored in this format, you'll need to use *all_labels.csv* to train the model to recognize each image's label. This is where the flow_from_dataframe method comes in.

Downloading Images and Setting Up Directories

Let's start with an example in which images are organized into a single directory. Download the file (*https://oreil.ly/WtKvA*) *flower_photos.zip*, unzip it, and you will see the directory structure shown in Figure 5-1.

Alternatively, if you're working in a Jupyter Notebook environment, run the Linux command wget to download *flower_photos.zip*. Following is the command for a Jupyter Notebook's cell:

```
!wget https://data.mendeley.com/public-files/datasets/jxmfrvhpyz/
files/283400ff-e529-4c3c-a1ee-4fb90024dc94/file_downloaded \
--output-document flower_photos.zip
```

The preceding command downloads the file and places it in the current directory. Unzip the file with this Linux command:

```
!unzip -q flower_photos.zip
```

This creates a directory with the same name as the ZIP file:

```
drwxr-xr-x 3 root root      4096 Nov  9 03:24 flower_photos
-rw-r--r-- 1 root root 228396554 Nov  9 20:14 flower_photos.zip
```

As you can see, there is a directory named *flower_photos*. List its contents with the following command, and you will see exactly what's shown in Figure 5-1:

```
!ls -alt flower_photos
```

Now that you have the directory structure and image files you need to work through this section's example, you can start building a data pipeline to feed these images into an image classification model for training. And to make it easy on yourself, you'll use the ResNet feature vector, a prebuilt model in TensorFlow Hub, so you don't have to design a model,. You'll stream these images into the training process with `ImageData Generator`.

Creating the Data Ingestion Pipeline

As usual, the first thing to do is import the necessary libraries:

```
import tensorflow as tf
import tensorflow_hub as hub
import pandas as pd
import numpy as np
import matplotlib.pyplot as plt
```

Notice that you need the pandas library in this example. This library is used to parse the label files as a dataframe. This is how to read the label file into a pandas DataFrame:

```
traindf=pd.read_csv('flower_photos/all_labels.csv',dtype=str)
```

And if you take a look at the dataframe `traindf`, you will see the following content.

	Filename	Label
0	7176723954_e41618edc1_n.jpg	sunflowers
1	2788276815_8f730bd942.jpg	roses
2	6103898045_e066cdeedf_n.jpg	dandelion

	Filename	Label
3	1441939151_b271408c8d_n.jpg	daisy
4	2491600761_7e9d6776e8_m.jpg	roses
...	...	...
3615	9558628596_722c29ec60_m.jpg	sunflowers
3616	4580206494_9386c81ed8_n.jpg	tulips

Next, you need to create some variables to hold parameters to be used later:

```
data_root = 'flower_photos/flowers'
IMAGE_SIZE = (224, 224)
TRAINING_DATA_DIR = str(data_root)
BATCH_SIZE = 32
```

Also, remember that when we use the ResNet feature vector, we have to rescale the image pixel intensity to a range of [0, 1], which means for each image pixel, the intensity has to be divided by 255. Also, we need to reserve a portion of the images for cross validation—say, 20%. So let's define these criteria in a dictionary, which we can use as an input for our ImageDataGenerator definition:

```
datagen_kwargs = dict(rescale=1./255, validation_split=.20)
```

Another dictionary will hold a few other arguments. The ResNet feature vector expects images to have pixel dimensions of 224 × 224, and we need to specify the batch size and resample algorithm as well:

```
dataflow_kwargs = dict(target_size=IMAGE_SIZE,
batch_size=BATCH_SIZE,
interpolation="bilinear")
```

This dictionary will be used as an input in the data flow definition.

For training the images, this is how you would set up the generator definition:

```
train_datagen = tf.keras.preprocessing.image.
                ImageDataGenerator(**datagen_kwargs)
```

Notice that we passed `datagen_kwargs` into the `ImageDataGener`
ator instance. Next, we use the `flow_from_dataframe` method
to create a data flow pipeline:

```
train_generator=train_datagen.flow_from_dataframe(
dataframe=traindf,
directory=data_root,
x_col="file_name",
y_col="label",
subset="training",
seed=10,
shuffle=True,
class_mode="categorical",
**dataflow_kwargs)
```

The `ImageDataGenerator` we defined as `train_datagen` is used
to invoke the `flow_from_dataframe` method. Let's take a look a
the input parameters. The first argument is `dataframe`, which is
designated as `traindf`. Then `directory` specifies where images
may be found in the directory path. `x_col` and `y_col` are the
headers in `traindf`: `x_col` corresponds to column "file_name"
as defined in *all_labels.csv*, and `y_col` is the column "label."
Now our generator knows how to match images to their labels.

Next, it specifies a subset to be `training`, as this is the training
image generator. Seed is provided for reproducibility of
batches. Images are shuffled, and image classes are designated
to be categorical. Finally, `dataflow_kwargs` is passed into this
`flow_from_dataframe` method so that raw images are resampled
from their original resolution to 224 × 224 pixels.

This procedure is repeated for the validation image generator:

```
valid_datagen = tf.keras.preprocessing.image.ImageDataGenerator(
**datagen_kwargs)
valid_generator=valid_datagen.flow_from_dataframe(
dataframe=traindf,
directory=data_root,
x_col="file_name",
y_col="label",
subset="validation",
seed=10,
shuffle=True,
class_mode="categorical",
**dataflow_kwargs)
```

Inspecting the Dataset

Right now, the only way to examine the contents of a Tensor-Flow dataset is by iterating through it:

```
image_batch, label_batch = next(iter(train_generator))
fig, axes = plt.subplots(8, 4, figsize=(20, 40))
axes = axes.flatten()
for img, lbl, ax in zip(image_batch, label_batch, axes):
    ax.imshow(img)
    label_ = np.argmax(lbl)
    label = idx_labels[label_]
    ax.set_title(label)
    ax.axis('off')
plt.show()
```

The preceding code snippet acquires the first batch of images from train_generator, the output of which is a tuple consisting of image_batch and label_batch.

You will see 32 images (that's the batch size). Some will look like Figure 5-2.

Figure 5-2. Some of the flower images in the dataset

Now that the data ingestion pipeline is set up, you are ready to use it in the training process.

Building and Training the tf.keras Model

The following classification model is an example of how to use a prebuilt model in TensorFlow Hub:

```
mdl = tf.keras.Sequential([
      tf.keras.layers.InputLayer(input_shape=IMAGE_SIZE + (3,)),
                  hub.KerasLayer(
"https://tfhub.dev/tensorflow/resnet_50/feature_vector/1",
trainable=False),

tf.keras.layers.Dense(5, activation='softmax',
name = 'custom_class')
])
mdl.build([None, 224, 224, 3])
```

Once the model architecture is ready, compile it:

```
mdl.compile(
  optimizer=tf.keras.optimizers.SGD(lr=0.005, momentum=0.9),
  loss=tf.keras.losses.CategoricalCrossentropy(
  from_logits=True,
  label_smoothing=0.1),
  metrics=['accuracy'])
```

Then launch the training process:

```
steps_per_epoch = train_generator.samples //
train_generator.batch_size
validation_steps = valid_generator.samples //
valid_generator.batch_size

mdl.fit(
    train_generator,
    epochs=13, steps_per_epoch=steps_per_epoch,
    validation_data=valid_generator,
    validation_steps=validation_steps)
```

Notice that `train_generator` and `valid_generator` are passed into our `fit` function. These will generate samples of images as the training process progresses, until all epochs are completed. You should expect to see output similar to this:

```
Epoch 10/13
90/90 [==============================] - 17s 194ms/step
loss: 1.0338 - accuracy: 0.9602 - val_loss: 1.0779
val_accuracy: 0.9020
Epoch 11/13
90/90 [==============================] - 17s 194ms/step
```

```
loss: 1.0311 - accuracy: 0.9623 - val_loss: 1.0750
val_accuracy: 0.9077
Epoch 12/13
90/90 [==============================] - 17s 193ms/step
loss: 1.0289 - accuracy: 0.9672 - val_loss: 1.0741
val_accuracy: 0.9091
Epoch 13/13
90/90 [==============================] - 17s 192ms/step
loss: 1.0266 - accuracy: 0.9693 - val_loss: 1.0728
val_accuracy: 0.9034
```

This indicates that you've successfully passed the training image generator and validation image generator into the training process, and that both generators can ingest data at training time. The result for validation data accuracy, val_accuracy, is a good indication that our choice of the ResNet feature vector works well for our use case of classifying flower images.

Streaming a NumPy Array with the from_tensor_slices Method

You can also create a data pipeline that streams a NumPy array. You *could* pass a NumPy array into the model training process directly, but to utilize RAM and other system resources efficiently, it's better to build a data pipeline. Further, once you are happy with the model and are ready to scale it up to handle a larger volume of data in production, you'll need to have a data pipeline anyway. Therefore, it is a good idea to build one, even for simple data structures such as a NumPy array.

Python's NumPy array is a versatile data structure. It can be used to represent numeric vectors and tabular data as well as raw images. In this section, you will learn how to use the from_tensor_slices method to stream NumPy data as a dataset.

The example NumPy data you will use for this section is the Fashion-MNIST dataset (*https://oreil.ly/CaUbq*), which consists of 10 types of garments in grayscale images. The images are represented using a NumPy structure instead of a typical image format, such as JPEG or PNG. There are 70,000 images in total.

The dataset is available in TensorFlow's distribution and can be easily loaded using the `tf.keras` API.

Loading Example Data and Libraries

To start, let's load the necessary libraries and the Fashion-MNIST data:

```
import tensorflow as tf
import numpy as np
import matplotlib.pyplot as plt

fashion_mnist = tf.keras.datasets.fashion_mnist
(train_images, train_labels),
(test_images, test_labels) = fashion_mnist.load_data()
```

This data is loaded with the `load_data` function in the `tf.keras` API. The data is partitioned into two tuples. Each tuple consists of two NumPy arrays, images and labels, as confirmed by the following command:

```
print(type(train_images), type(train_labels))

<class 'numpy.ndarray'> <class 'numpy.ndarray'>
```

This confirms the data type. It is important to know the array dimension, which you can display using the `shape` command:

```
print(train_images.shape, train_labels.shape)

(60000, 28, 28) (60000,)
```

As you can see, `train_images` is made up of 60,000 records, each a 28 × 28 NumPy array, while `train_labels` is a 60,000-record label index. TensorFlow provides a useful tutorial (*https://oreil.ly/7d85v*) on how these indices map to class names, but here is a quick look.

Label	Class
0	T-shirt/top
1	Trouser
2	Pullover

Label	Class
3	Dress
4	Coat
5	Sandal
6	Shirt
7	Sneaker
8	Bag
9	Ankle boot

Inspecting the NumPy Array

Next, inspect one of the records to see the images for yourself. To display a NumPy array as a color scale, you'll need to use the matplotlib library, which you imported earlier. The object plt represents this library:

```
plt.figure()
plt.imshow(train_images[5])
plt.colorbar()
plt.grid(False)
plt.show()
```

Figure 5-3 displays the NumPy array for train_images[5].

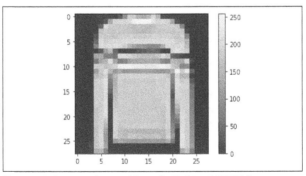

Figure 5-3. An example record from the Fashion-MNIST dataset

Unlike color images in JPEG format, which contain three separate channels (RGB), each image in the Fashion-MNIST dataset is represented as a flat, two-dimensional structure of 28 × 28 pixels. Notice that the pixel values are between 0 and 255; we need to normalize them to [0, 1].

Building the Input Pipeline for NumPy Data

Now you are ready to build a streaming pipeline. First you need to normalize each pixel in the image to within the range [0, 1]:

```
train_images = train_images/255
```

Now the data value is correct, and it is ready to be passed to the `from_tensor_slices` method:

```
train_dataset = tf.data.Dataset.from_tensor_slices((train_images,
train_labels))
```

Next, split this dataset into training and validation sets. In the following code snippet, I specify that the validation set is 10,000 images, with the remaining 50,000 images going into the training set:

```
SHUFFLE_BUFFER_SIZE = 10000
TRAIN_BATCH_SIZE = 50
VALIDATION_BATCH_SIZE = 10000

validation_ds = train_dataset.shuffle(SHUFFLE_BUFFER_SIZE).
take(VALIDATION_SAMPLE_SIZE).
batch(VALIDATION_BATCH_SIZE)

train_ds = train_dataset.skip(VALIDATION_BATCH_SIZE).
batch(TRAIN_BATCH_SIZE).repeat()
```

When cross validation is part of the training process, you also need to define a couple of parameters so that the model knows when to stop and evaluate the cross-validation data during the training iteration:

```
steps_per_epoch = 50000 // TRAIN_BATCH_SIZE
validation_steps = 10000 // VALIDATION_BATCH_SIZE
```

The following is a small classification model:

```
model = tf.keras.Sequential([
    tf.keras.layers.Flatten(input_shape=(28, 28)),
    tf.keras.layers.Dense(30, activation='relu'),
    tf.keras.layers.Dense(10)
])

model.compile(optimizer=tf.keras.optimizers.RMSprop(),
  loss=tf.keras.losses.SparseCategoricalCrossentropy(
  from_logits=True),
  metrics=['sparse_categorical_accuracy'])
```

Now you can start the training:

```
model.fit(
    train_ds,
    epochs=13, steps_per_epoch=steps_per_epoch,
    validation_data=validation_ds,
    validation_steps=validation_steps)
```

Your output should look similar to this:

```
...
Epoch 10/13
1562/1562 [==============================] - 4s 3ms/step
loss: 0.2982 - sparse_categorical_accuracy: 0.8931
val_loss: 0.3476 - val_sparse_categorical_accuracy: 0.8778
Epoch 11/13
1562/1562 [==============================] - 4s 3ms/step
loss: 0.2923 - sparse_categorical_accuracy: 0.8954
val_loss: 0.3431 - val_sparse_categorical_accuracy: 0.8831
Epoch 12/13
1562/1562 [==============================] - 4s 3ms/step
loss: 0.2867 - sparse_categorical_accuracy: 0.8990
val_loss: 0.3385 - val_sparse_categorical_accuracy: 0.8854
Epoch 13/13
1562/1562 [==============================] - 4s 3ms/step
loss: 0.2826 - sparse_categorical_accuracy: 0.8997
val_loss: 0.3553 - val_sparse_categorical_accuracy: 0.8811
```

Notice that you can pass train_ds and validation_ds to the fit function directly. This is exactly the same method you learned in Chapter 4, when you built an image generator and trained the image classification model to classify five types of flowers.

Wrapping Up

In this chapter, you learned how to build data pipelines for text, numeric arrays, and images. As you have seen, data and directory structure are important to set up before applying different APIs to ingest the data to a model. We started with a text data example, using the `text_dataset_from_directory` function that TensorFlow provides to handle text files. You also learned that the `flow_from_dataframe` method is specifically designed for image files grouped by class, a totally different file structure than what you saw in Chapter 4. Finally, for numeric arrays in a NumPy array structure, you learned to use the `from_tensor_slices` method to build a dataset for streaming. When you build a data ingestion pipeline, you have to understand the file structure as well as the data type in order to use the right method.

Now that you have seen how to build data pipelines, you'll learn more about building the model in the next chapter.

## CHAPTER 6

Model Creation Styles

As you may have imagined, there is more than one way to build a deep learning model. In the previous chapters, you learned about tf.keras.Sequential, known as the *symbolic API*, which is commonly the starting point when teaching model creation. Another style of API that you might come across is known as the *imperative API*. Both symbolic and imperative APIs are capable of building deep learning models.

By and large, which API you choose is a matter of style. Depending on your programming experience and background, one or the other might feel more natural for you. In this chapter, you will learn how to build the same model with both APIs. Specifically, you will learn how to build an image classification model using the CIFAR-10 image dataset (*https://oreil.ly/ W81qK*). This dataset consists of 10 commonly seen *classes*, or categories, of images. Like the flower images we used previously, the CIFAR-10 images are available as part of the TensorFlow distribution. However, while the flower images came in JPEG format, the CIFAR-10 images are NumPy arrays. To stream them into the training process, instead of using the flow_from_directory method as you did in Chapter 5, you'll use the from_tensor_slices method.

After establishing the data streaming process with `from_ten sor_slices`, you'll first use the symbolic API to build and train the image classification model, and then use the imperative API. You will see that regardless of how you build the model architecture, the results are very similar.

Using the Symbolic API

You have already seen the symbolic API, `tf.keras.Sequential`, at work in this book's examples. Within `tf.keras.Sequential` there are stacks of layers, each of which performs certain operations to input data. Since models are built layer by layer, this is an intuitive way to envision the process. In most cases, you only have one source of input (in this case, a stream of images), and the output is the class of input images. In "Model Implementation with TensorFlow Hub" on page 74, you learned how to build a model with TensorFlow Hub. The model architecture is defined with the sequential API, as shown in Figure 6-1.

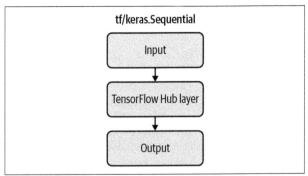

Figure 6-1. Sequential API pattern and data flow

In this section, you will learn how to use this API to build and train an image classification model with CIFAR-10 images.

Loading the CIFAR-10 Images

The CIFAR-10 image dataset contains 10 classes: airplanes, automobiles, birds, cats, deer, dogs, frogs, horses, ships, and trucks. All images are 32 × 32 pixels and colored with three channels (RGB).

Start by importing the necessary libraries:

```
import tensorflow as tf
from tensorflow.keras import datasets, layers, models
import numpy as np
import matplotlib.pylab as plt

(train_images, train_labels), (test_images, test_labels) =
datasets.cifar10.load_data()
```

This code downloads the CIFAR-10 images to your Python runtime, partitioned into training and test sets. You can verify the format with a type statement:

```
print(type(train_images))
```

The output will be a data type:

```
<class 'numpy.ndarray'>
```

It is also important to know the array's shape, which you can find with the following command:

```
print(train_images.shape, train_labels.shape)
```

Here are the array shapes for the images and labels, respectively:

```
(50000, 32, 32, 3) (50000, 1)
```

You can do the same for the test data:

```
print(test_images.shape, test_labels.shape)
```

You should get the following output:

```
(10000, 32, 32, 3) (10000, 1)
```

As you can see from the outputs, the CIFAR-10 dataset consists of 50,000 training images, each $32 \times 32 \times 3$ pixels. The accompanying 50,000 labels are a one-dimensional array of indices that denote image classes. Likewise, there are 10,000 test images with corresponding labels. The label indices correspond to the following names:

```
CLASS_NAMES = ['airplane', 'automobile', 'bird', 'cat',
               'deer', 'dog', 'frog', 'horse', 'ship', 'truck']
```

Thus, an index of 0 denotes the label "airplane," while an index of 9 denotes "truck."

Inspecting Label Distribution

Now it's time to find out the distribution of these classes and see some of the images. To find out how many samples each class has, look at the distribution of training labels by class using the NumPy `unique` function:

```
unique, counts = np.unique(train_labels, return_counts=True)
```

This will return sample counts for each label. To display it:

```
print(np.asarray((unique, counts)))
```

It will show the following:

```
[[   0    1    2    3    4    5    6    7    8    9]
 [5000 5000 5000 5000 5000 5000 5000 5000 5000 5000]]
```

This means that there are 5,000 images in each label (class). The training data is evenly distributed among all labels.

Similarly, you can verify the distribution of the test data:

```
unique, counts = np.unique(test_labels, return_counts=True)
print(np.asarray((unique, counts)))
```

The output confirms the count of 1,000 images for each label:

```
[[   0    1    2    3    4    5    6    7    8    9]
 [1000 1000 1000 1000 1000 1000 1000 1000 1000 1000]]
```

Inspecting Images

Let's take a look at some of the images to ensure their data quality. For this exercise, you'll randomly sample and display 25 images of the 50,000 in the training dataset.

How does TensorFlow make this random selection? The images are indexed from 0 to 49,999. To randomly select a finite number of indices from this range, use Python's `random` library, which takes a Python list as input and randomly selects a finite number of samples from it:

```
selected_elements = random.sample(a_list, 25)
```

This code randomly selects 25 elements from `a_list` and stores the results in `selected_elements`. If `a_list` corresponds to the image indices, then `selected_elements` will contain 25 indices drawn at random from `a_list`. You will use `selected_elements` to access and display these 25 training images.

Now you need to create `train_idx`, the list that holds the indices for training images. You'll use the Python `range` function to create an object that holds integers in the range 0 to 49,999:

```
range(len(train_labels))
```

The preceding code creates a `range` object that holds integers that start at 0 and go up to `len(train_labels)`, or the length of the list `training_labels`.

Now, convert the `range` object to a Python list:

```
list(range(len(train_labels)))
```

This list is now ready to serve as input to the Python `random.sample` function. Now you can start on your code.

First, create `train_idx`, which is a list of indices from 0 to 49,999:

```
train_idx = list(range(len(train_labels)))
```

Then use the `random` library to generate the random selection:

```
import random
random.seed(2)
random_sel = random.sample(train_idx, 25)
```

The seed operation in the second line ensures that your selection is reproducible, which is helpful for debugging purposes. You can use any integer for `seed`.

The `random_sel` list will hold 25 randomly selected indices that look something like this:

```
[3706,
 6002,
 5562,
 23662,
 11081,
 48232,
 ...
```

Now you can plot images based on these indices and display their labels:

```
plt.figure(figsize=(10,10))
for i in range(len(random_sel)):
 plt.subplot(5,5,i+1)
 plt.xticks([])
 plt.yticks([])
 plt.grid(False)
 plt.imshow(train_images[random_sel[i]], cmap=plt.cm.binary)
 plt.xlabel(CLASS_NAMES[train_labels[random_sel[i]][0]])
plt.show()
```

This code snippet displays a panel of 25 images along with their labels, as shown in Figure 6-2. (Because this is a random sample, your results will vary.)

Figure 6-2. Twenty-five images from the CIFAR-10 dataset, selected at random

Building a Data Pipeline

In this section, you will build a data ingestion pipeline using `from_tensor_slices`. Since there are only two partitions, training and test, you'll need to reserve half of the test partition for cross validation during the training process. Select the first 500 as cross-validation data and the remaining 500 as test data:

```
validation_dataset = tf.data.Dataset.from_tensor_slices(
(test_images[:500], test_labels[:500]))

test_dataset = tf.data.Dataset.from_tensor_slices(
(test_images[500:], test_labels[500:]))
```

This code creates two dataset objects based on image indices, `validation_dataset` and `test_dataset`, with 500 samples in each set.

Now create a similar dataset object for the training data:

```
train_dataset = tf.data.Dataset.from_tensor_slices(
(train_images, train_labels))
```

All training images are used here. You can confirm that by counting the samples in `train_dataset`:

```
train_dataset_size = len(list(train_dataset.as_numpy_iterator()))
print('Training data sample size: ', train_dataset_size)
```

This is the expected result:

```
Training data sample size: 50000
```

Batching the Dataset for Training

To finish setting up the data ingestion pipeline for training, you will need to divide the training data into batches. The size of a batch, or number of training samples, is the number required for the model training process to update the model weights and bias, and then move along one step to reduce the error gradient.

Batch your training data with the following code, which first shuffles the training dataset and then creates multiple batches of 200 samples each:

```
TRAIN_BATCH_SIZE = 200
train_dataset = train_dataset.shuffle(50000).batch(
TRAIN_BATCH_SIZE)
```

Likewise, you'll do the same for cross-validation and test data:

```
validation_dataset = validation_dataset.batch(500)
test_dataset = test_dataset.batch(500)

STEPS_PER_EPOCH = train_dataset_size // TRAIN_BATCH_SIZE
VALIDATION_STEPS = 1 #validation data // validation batch size
```

The cross-validation and test datasets each consist of one 500-sample batch. The code sets parameters to inform the training process how many batches of training and validation data to expect. The parameter for training data is STEPS_PER_EPOCH. The parameter for cross-validation data is VALIDATION_STEPS and it is set to 1 because the data size and batch size are both 500. Note that a double slash (//) denotes *floor division* (that is, rounding down to the nearest integer).

Now that your training and validation datasets are ready, your next step is to build the model with the symbolic API.

Building the Model

Now you are ready to build the model. Here is example code for a deep learning, image-classification model, built with a stack of layers wrapped by the tf.keras.Sequential class:

```
model = tf.keras.Sequential([
 tf.keras.layers.Conv2D(32, kernel_size=(3, 3), activation='relu',
 kernel_initializer='glorot_uniform', padding='same',
 input_shape = (32,32,3)),
 tf.keras.layers.MaxPooling2D(pool_size=(2, 2)),
 tf.keras.layers.Conv2D(64, kernel_size=(3, 3), activation='relu',
 kernel_initializer='glorot_uniform',
 padding='same'),
 tf.keras.layers.MaxPooling2D(pool_size=(2, 2)),
 tf.keras.layers.Flatten(),
 tf.keras.layers.Dense(256, activation='relu',
 kernel_initializer='glorot_uniform'),
 tf.keras.layers.Dense(10, activation='softmax',
 name = 'custom_class')
])
model.build([None, 32, 32, 3])
```

Next, compile the model with the loss function designated for a classification task:

```
model.compile(
 loss='sparse_categorical_crossentropy',
 optimizer=tf.keras.optimizers.SGD(learning_rate=0.1,
 momentum=0.9),
 metrics=['accuracy'])
```

To envision how the model handles and transforms data through different layers, you may wish to plot the model architecture, including the input and output shapes of the tensors it expects. You can use this command:

```
tf.keras.utils.plot_model(model, show_shapes=True)
```

You might need to install the `pydot` and `graphviz` libraries before running this command:

```
pip install pydot
pip install graphviz
```

Figure 6-3 shows the model architecture. The question marks indicate the dimension that denotes sample size, which is only known during execution. This is because the model is designed to work with training samples of any size. The memory required to handle sample size is irrelevant and need not be specified at the model architecture level. Instead, the required memory will be defined during the training execution.

Next, start the training process:

```
hist = model.fit(
 train_dataset,
 epochs=5, steps_per_epoch=STEPS_PER_EPOCH,
 validation_data=validation_dataset,
 validation_steps=VALIDATION_STEPS).history
```

Your results should be similar to those in Figure 6-4.

This is how to leverage `tf.keras.Sequential` to build and train a deep learning model. As you can see, as long as you specify the input and output shapes to be consistent with the images and labels, you can stack as many layers as you wish. The training process is also pretty routine; it doesn't deviate from what you saw in Chapter 5.

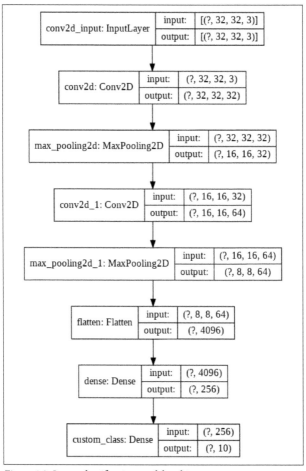

Figure 6-3. Image classification model architecture

```
Epoch 1/5
250/250 [==============================] - 73s 251ms/step - loss: 1.3419 - accuracy: 0.3239 - val_loss: 1.4267 - val_accuracy: 0.4320
Epoch 2/5
250/250 [==============================] - 72s 289ms/step - loss: 1.3386 - accuracy: 0.5255 - val_loss: 1.3181 - val_accuracy: 0.5020
Epoch 3/5
250/250 [==============================] - 72s 289ms/step - loss: 1.1898 - accuracy: 0.5781 - val_loss: 1.2002 - val_accuracy: 0.5520
Epoch 4/5
250/250 [==============================] - 72s 290ms/step - loss: 1.0525 - accuracy: 0.6288 - val_loss: 1.1291 - val_accuracy: 0.6280
Epoch 5/5
250/250 [==============================] - 73s 292ms/step - loss: 0.9388 - accuracy: 0.6701 - val_loss: 1.1946 - val_accuracy: 0.5940
```

Figure 6-4. Model training results

Before we look at the imperative API, we're going to take a quick detour: you'll need to know a bit about class inheritance in Python to understand the imperative API.

Understanding Inheritance

Inheritance is a technique used in object-oriented programming. It uses the concept of *classes* to encapsulate attributes and methods associated with a particular type of object. It also handles relationships between different types of objects. Inheritance is the means by which a particular class is allowed to use methods in another class.

It is much easier to see how this works with a trivial example. Imagine we have a base (or parent) class, called vehicle. We also have another class, truck, which is a *child class* of vehicle: this is also known as a *derived class* or *inherited class*. We can define the vehicle class as follows:

```python
class vehicle():
 def __init__(self, make, model, horsepower, weight):
 self.make = make
 self.model = model
 self.horsepower = horsepower
 self.weight = weight

 def horsepower_to_weight_ratio(self, horsepower, weight):
 hp_2_weight_ratio = horsepower / weight
 return hp_2_weight_ratio
```

This code shows a common pattern for defining a class. It has a constructor, __init__, which initializes the class's attributes, such as make, model, horsepower, and weight. Then there is a function called horsepower_to_weight_ratio which, as you probably gathered, calculates the horsepower-to-weight ratio of a vehicle (we'll call this the HW ratio). This function is also accessible to any child class of the vehicle class.

Now let's create truck, the child class for vehicle:

```python
class truck(vehicle):
 def __init__(self, make, model, horsepower, weight, payload):
 super().__init__(make, model, horsepower, weight)
 self.payload = payload

 def __call__(self, horsepower, payload):
 hp_2_payload_ratio = horsepower / payload
 return hp_2_payload_ratio
```

In this definition, class truck(vehicle) indicates that truck is a child class of vehicle.

In the constructor __init__, super returns a temporary object of the parent class vehicle to the truck class. This object then calls the parent class's __init__, which enables the truck class to reuse the same attributes defined in the parent class: make, model, horsepower, and weight. However, a truck also has an attribute that is unique: payload. This attribute is *not* inherited from the base class; rather, it is defined in the truck class. You can define payload with self.payload = payload. Here, the self keyword refers to the instance of this class. In this case, it is a truck instance, and payload is an arbitrary name you defined for this attribute.

Next, there is a __call__ function. This function makes the truck class "callable." Before we look into what __call__ does or what it means for a class to be callable, let's define a few parameters and create a truck instance:

```python
MAKE = 'Tesla'
MODEL = 'Cybertruck'
HORSEPOWER = 800 #HP
WEIGHT = 3000 #kg
PAYLOAD = 1600 #kg

MyTruck = truck(MAKE, MODEL, HORSEPOWER, WEIGHT, PAYLOAD)
```

To make sure this has been done properly, print these attributes:

```
print('Make: ', MyTruck.make,
 '\nModel: ', MyTruck.model,
 '\nHorsepower (HP): ', MyTruck.horsepower,
 '\nWeight (kg): ', MyTruck.weight,
 '\nPayload (kg): ', MyTruck.payload)
```

This should produce the following output:

```
Make: Tesla
Model: Cybertruck
Horsepower (HP): 800
Weight (kg): 3000
Payload (kg): 1600
```

What does it mean to make a Python class *callable*? Let's say that you're a bricklayer and you need to haul heavy loads in your truck. For you, the most important attribute of a truck is its horsepower-to-payload ratio (HP ratio). Fortunately, you can create an instance of the truck object and calculate the ratio right away:

```
MyTruck(HORSEPOWER, PAYLOAD)
```

The output will be 0.5.

This means that the MyTruck instance actually has a value associated with it. This value is defined as the horsepower-to-payload ratio. The calculation is done by the __call__ function of the truck class, which is a built-in function for Python classes. When this function is explicitly defined to perform a certain logic, it works almost like a function call. Look at this line of code again:

```
MyTruck(HORSEPOWER, PAYLOAD)
```

If you saw only this line, you might well think that MyTruck is a function, and HORSEPOWER and PAYLOAD are the inputs.

By explicitly defining the __call__ method to calculate the HP ratio, you made the truck class callable; in other words, you made it behave like a function. Now it can be called like a Python function.

Next we want to find the HW ratio of our object `MyTruck`. You may notice that no method for this is defined in the `truck` class. However, since there *is* such a method in the parent `vehicle` class, `horsepower_to_weight_ratio`, `MyTruck` can perform the calculation using this method. This is a demonstration of *class inheritance*, where a child class may use a method defined directly by the parent class. To do this, you'd use:

```
MyTruck.horsepower_to_weight_ratio(HORSEPOWER, WEIGHT)
```

The output is 0.26666666666666666.

Using the Imperative API

Having seen how Python's class inheritance works, you are now ready to learn how to use the imperative API to build a model. The imperative API is also known as the *model subclassing API* because any model you build is really inherited from a "Model" class. If you are familiar with an object-oriented programming language such as C#, C++, or Java, the imperative style should feel familiar.

Defining a Model as a Class

How would you define the model you built in the preceding section as a class? Let's look at the code:

```
class myModel(tf.keras.Model):
 def __init__(self, input_dim):
 super(myModel, self).__init__()
 self.conv2d_initial = tf.keras.layers.Conv2D(32,
 kernel_size=(3, 3),
 activation='relu',
 kernel_initializer='glorot_uniform',
 padding='same',
 input_shape = (input_dim,input_dim,3))
 self.cov2d_mid = tf.keras.layers.Conv2D(64, kernel_size=(3, 3),
 activation='relu',
 kernel_initializer='glorot_uniform',
 padding='same')
 self.maxpool2d = tf.keras.layers.MaxPooling2D(pool_size=(2, 2))
 self.flatten = tf.keras.layers.Flatten()
 self.dense = tf.keras.layers.Dense(256, activation='relu',
 kernel_initializer='glorot_uniform')
```

```
self.fc = tf.keras.layers.Dense(10, activation='softmax',
name = 'custom_class')

def call(self, input_dim):
x = self.conv2d_initial(input_dim)
x = self.maxpool2d(x)
x = self.cov2d_mid(x)
x = self.maxpool2d(x)
x = self.flatten(x)
x = self.dense(x)
x = self.fc(x)

return x
```

As the preceding code indicates, the myModel class inherits from the parent class tf.keras.Model in exactly the same way our truck class inherits from the parent class vehicle.

The layers in the model are treated as attributes in the myModel class. These attributes are defined in the constructor function __init__. (Recall that attributes are parameters, such as horsepower, make, and model, while layers are defined by syntax, such as tf.keras.layers.Conv2D.) For the first layer in the model, the code is:

```
self.conv2d_initial = tf.keras.layers.Conv2D(32,
 kernel_size=(3, 3),
 activation='relu',
 kernel_initializer='glorot_uniform',
 padding='same',
 input_shape = (input_dim,input_dim,3))
```

As you can see, all it takes to assign the layer is an object named conv2d_initial. Another important element in this definition is that you can pass a user-defined parameter into an attribute. Here, the constructor function __init__ expects the user to provide an argument, input_dim, which it will pass to the input_shape argument.

The benefit of this style is that if you want to reuse this model architecture for other types of image dimensions, you don't have to create a new model; you just pass the image dimension as a user argument to this class, and you will get an instance of the class that can handle the image dimensions of your choice.

In fact, you can add more user arguments to the input of the constructor function and pass them into different parts of the object, such as `kernel_size`. This is one way the object-oriented programming style promotes code reuse.

Let's take a look at another layer definition:

```
self.maxpool2d = tf.keras.layers.MaxPooling2D(pool_size=(2, 2))
```

This layer will be used exactly as is multiple times in the model architecture, but you need to define it only once. However, if you need a different hyperparameter value, such as a different `pool_size`, then you need to create another attribute:

```
self.maxpool2d_2 = tf.keras.layers.MaxPooling2D(pool_size=(5, 5))
```

Here, there is no need to do so, since our model architecture reuses `maxpool2d` as is.

Now let's take a look at the `call` function. Recall that you make a class callable by handling certain types of logic or calculations in Python's built-in `__call__` function. In a similar spirit, TensorFlow created a built-in `call` function that makes the model class callable. Inside this function, you can see that the order of layers is the same as what goes into the sequential API (as you saw in "Building the Model" on page 111. The only difference is that these layers are now represented by class attributes, instead of a hardcoded layer definition.

Also, notice that in the following input, the user argument `input_dim` is passed into the attributes:

```
def call(self, input_dim)
```

This can provide flexibility and reusability to your model, depending on your image dimension requirements.

In the `call` function, the object x is used to represent the model layer iteratively. After the final layer, `self.fc(x)`, is declared, it returns x as the model.

To create an instance of a model that handles CIFAR-10's image dimension of 32 × 32 pixels, define the instance as:

```
mdl = myModel(32)
```

This code creates an instance of `myModel` and initializes it with the CIFAR-10 dataset's image dimension. This model is represented as the `mdl` object. Next, just as you did in "Building the Model" on page 111, you have to designate a loss function and optimizer choice with the same syntax:

```
mdl.compile(loss='sparse_categorical_crossentropy',
 optimizer=tf.keras.optimizers.SGD(learning_rate=0.1,
 momentum=0.9),
 metrics=['accuracy'])
```

Now you may launch the training routine:

```
mdl_hist = mdl.fit(
 train_dataset,
 epochs=5, steps_per_epoch=STEPS_PER_EPOCH,
 validation_data=validation_dataset,
 validation_steps=VALIDATION_STEPS).history
```

You can expect a training outcome similar to the one in Figure 6-5.

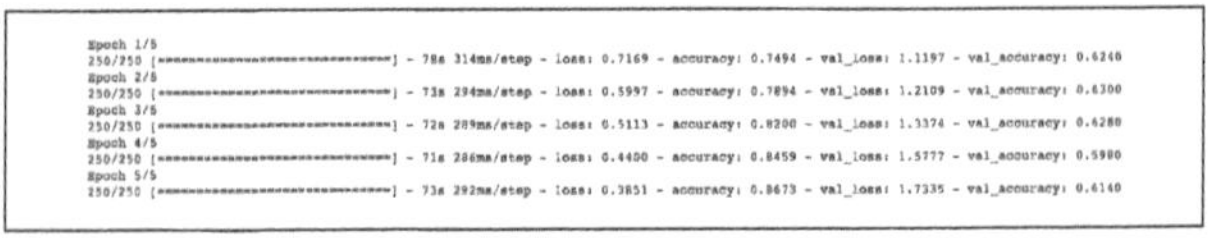

Figure 6-5. Imperative API model training results

Models trained with the symbolic API and the imperative API should produce similar training results.

Choosing the API

You have seen that the symbolic and imperative APIs can be used to build models with the same architecture. In most instances, your choice of API will be based on your preferred style and your familiarity with the syntax. However, there are some trade-offs that are worth noting.

The biggest advantage of the symbolic API is its code readability, which makes maintenance easier. It is straightforward to see the model architecture, and you can see the input data as a flow of tensors through different layers, like a graph. Models built with the symbolic API can also leverage tf.keras.utils.plot_model to show the model architecture. Typically, this is where most of us would start when designing a deep learning model.

The imperative API is definitely not as straightforward as the symbolic API when it comes to implementing a model architecture. As you've learned, this style stems from the object-oriented programming technique of class inheritance. If you are more comfortable with envisioning a model as an object rather than as a stack of operation layers, you may find this style more intuitive, as shown in Figure 6-6.

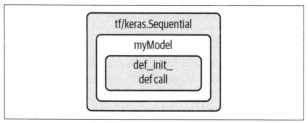

Figure 6-6. The TensorFlow model's imperative API (a.k.a. model subclassing)

Essentially, any model you build is an *extension*, or inherited class, of the base model tf.keras.Model. Thus, when you build a model, you are really just creating an instance of a class that inherits all the attributes and functions from that base model. To fit the model for images of different dimensions, you simply have to instantiate it with different hyperparameters. If reusing the same model architecture is a part of your workflow, then the imperative API is a sensible choice to keep your code clean and concise.

Using the Built-In Training Loop

So far, you have seen that all it takes to launch the model training process is the `fit` function. This function wraps a lot of complex operations for you, as Figure 6-7 shows.

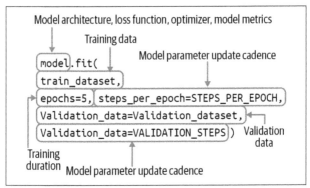

Figure 6-7. Elements in a built-in training loop

The model object contains information about the architecture, loss function, optimizer, and model metrics. Inside `fit`, you provide the training and validation data, the number of epochs to train it for, and how often to update the model parameters and test them with validation data.

That's all you have to do. The built-in training loop knows that when an epoch of training is complete, it's time to perform cross validation with batched validation data. This is convenient and clear, and makes your code very easy to maintain. The output is produced at the end of each epoch, as seen in Figures 6-4 and 6-5.

If you ever need to look into the details of the training process, such as model accuracy within each step of incremental improvement before an epoch reaches the end, or if you ever want to create your own training metrics, then you need to build your own training loop. Next, we'll look at how that works.

Creating and Using a Custom Training Loop

With custom training loops, you lose the convenience of the fit function; instead, you have to write code to orchestrate the training process. Let's say you want to monitor accuracy during each step of the model parameter within an epoch. You can reuse the model object (model) from "Building the Model" on page 111.

Creating the Elements of the Loop

First, create the optimizer and loss function objects:

```
optimizer = tf.keras.optimizers.SGD(
learning_rate=0.1,
 momentum=0.9)
loss_fn = tf.keras.losses.SparseCategoricalCrossentropy(
from_logits=True)
```

Then create objects to represent model metrics:

```
train_acc_metric = tf.keras.metrics.SparseCategoricalAccuracy()
val_acc_metric = tf.keras.metrics.SparseCategoricalAccuracy()
```

This code creates two objects for model accuracy: one for the training data and one for the validation data. The SparseCate goricalAccuracy function is used because the output is a metric that calculates how often predictions match labels.

Then, for training, you need to create a function:

```
@tf.function
def train_step(train_data, train_label):
    with tf.GradientTape() as tape:
        logits = model(train_data, training=True)
        loss_value = loss_fn(train_label, logits)
        grads = tape.gradient(loss_value, model.trainable_weights)
        optimizer.apply_gradients(zip(grads, model.trainable_weights))
        train_acc_metric.update_state(train_label, logits)
        return loss_value
```

In the preceding code, @tf.function is a Python decorator that converts a function which takes a tensor as input. It helps speed up the function's execution. This function also includes a new object, tf.GradientTape. In this scope, TensorFlow executes the

---

gradient descent algorithm for you; it automatically calculates the gradient by differentiating the loss function with respect to training weights in each node.

The following line indicates the scope of the GradientTape object:

```
with tf.GradientTape() as tape
```

And this next line of code means that you call on model to map the training data to an output (logits):

```
logits = model(train_data, training=True)
```

Now calculate the output of the loss function when comparing the model output with the true label, train_label:

```
loss_value = loss_fn(train_label, logits)
```

Then use the model's parameters (trainable_weights) and the loss function's value (loss_value) to calculate the gradient and update the model's accuracy.

You'll need to do the same for the validation data:

```
@tf.function
def test_step(validation_data, validation_label):
 val_logits = model(validation_data, training=False)
 val_acc_metric.update_state(validation_label, val_logits)
```

Putting the Elements Together in a Custom Training Loop

Now that you have all the pieces, you're ready to create the custom training loop. Here is the general procedure:

1. Use a for loop to iterate through each epoch.

2. Within each epoch, use another for loop to iterate through each batch in the dataset.

3. In each batch, open a GradientTape object scope.

4. In the scope, compute the loss function.

5. Outside the scope, retrieve gradients of the model weights.

6. Use the optimizer to update the model weights based on the gradient values.

Following is the code snippet for a custom training loop:

```
import time

epochs = 2
for epoch in range(epochs):
 print("\nStarting epoch %d" % (epoch,))
 start_time = time.time()

 # Iterate dataset batches
 for step, (x_batch_train, y_batch_train) in
 enumerate(train_dataset):
 loss_value = train_step(x_batch_train, y_batch_train)

    # In every 100 batches, log results.
    if step % 100 == 0:
        print(
        "Training loss (for one batch) at step %d: %.4f"
        % (step, float(loss_value))
        )
 print("Sample processed so far: %d samples" %
 ((step + 1) * TRAIN_BATCH_SIZE))

 # Show accuracy metrics after each epoch is completed
 train_accuracy = train_acc_metric.result()
 print("Training accuracy over epoch: %.4f" %
 (float(train_accuracy),))

 # Reset training metrics before next epoch starts
 train_acc_metric.reset_states()

 # Test with validation data at end of each epoch
 for x_batch_val, y_batch_val in validation_dataset:
 test_step(x_batch_val, y_batch_val)

 val_accuracy = val_acc_metric.result()
 val_acc_metric.reset_states()
 print("Validation accuracy: %.4f" % (float(val_accuracy),))
 print("Time taken: %.2fs" % (time.time() - start_time))
```

Figure 6-8 shows typical output of the custom training loop execution.

```
Starting epoch 0
Training loss (for one batch) at step 0: 1.7271
Sample processed so far: 200 samples
Training loss (for one batch) at step 100: 1.7481
Sample processed so far: 20200 samples
Training loss (for one batch) at step 200: 1.7973
Sample processed so far: 40200 samples
Training accuracy over epoch: 0.6968
Validation accuracy: 0.6200
Time taken: 78.21s

Starting epoch 1
Training loss (for one batch) at step 0: 1.7332
Sample processed so far: 200 samples
Training loss (for one batch) at step 100: 1.7569
Sample processed so far: 20200 samples
Training loss (for one batch) at step 200: 1.7648
Sample processed so far: 40200 samples
Training accuracy over epoch: 0.7028
Validation accuracy: 0.6240
Time taken: 77.63s
```

Figure 6-8. Output from executing the custom training loop

As you can see, at the end of every batch of 200 samples, the training loop calculates and shows the loss function's value, giving you a microscopic view of what's going on inside the training process. If you need that kind of visibility, building your own custom training loop will provide it. Just know that it takes considerably more effort than the convenient built-in training loop of the fit function.

Wrapping Up

In this chapter, you learned how to build a deep learning model in TensorFlow using symbolic and imperative APIs. Very often, both are capable of accomplishing the same architecture, especially when the data flows from input to output in a straight line (meaning there are no feedbacks or multiple inputs). You may see models with complex architecture and customized implementations using the imperative API. Choose the API that suits your case, convenience, and readability.

Whichever you choose, you'll train the model in the same manner with the built-in `fit` function. The `fit` function executes the built-in training loop and shields you from worrying about how to actually orchestrate the training process. The details, such as calculating the loss function, comparing the model output with the true label, and updating the model parameter using the value of gradients, all happen behind the scenes for you. What you will see is the result at the end of each epoch: how accurate the model is with respect to training data and cross-validation data.

If you ever need a microscopic view of what's going on inside an epoch, such as how accurate the model is with each batch of training data, then you need to write your own training loop, which is a considerably more laborious process.

In the next chapter, you will see other options available within the model training process that provide even more flexibility, without the nuanced coding process of a custom training loop.

## CHAPTER 7

Monitoring the Training Process

In the last chapter, you learned how to launch the model training process. In this chapter, we'll cover the process itself.

I've used fairly straightforward examples in this book to help you grasp each concept. When you're running a real training process in TensorFlow, however, things can be more complicated. When problems arise, for example, you need to think about how to determine whether your model is *overfitting* the training data. (Overfitting occurs when the model learns and memorizes the training data and the noise in training data so well that it negatively affects its ability to learn new data.) If it is, you'll need to set up cross validation. If not, you can take steps to prevent overfitting.

Other questions that often arise during the training process include:

- How often should I save the model during the training process?

- How should I determine which epoch gives the best model before overfitting occurs?

- How can I track model performance?

- Can I stop training if the model is not improving or is overfitting?
- Is there a way to visualize the model training process?

TensorFlow provides a very easy way to address these questions: callback functions. In this chapter, you will learn how to make quick use of callback functions as you monitor the training process. The first half of the chapter discusses `ModelCheckpoint` and `EarlyStopping`, while the second half focuses on TensorBoard and shows you several techniques for invoking TensorBoard and using it for visualization.

Callback Objects

A TensorFlow *callback object* is an object that can perform a group of built-in functions provided by `tf.keras`. When certain events occur during training, a callback object will execute specific code or functions.

Using callbacks is optional, so you don't need to implement any callback objects to train a model. We'll be looking at three of the most frequently used classes: `ModelCheckpoint`, `EarlyStopping`, and TensorBoard.[1]

ModelCheckpoint

The `ModelCheckpoint` class enables you to save your model regularly throughout the training process. By default, at the end of each training epoch, model weights and biases are finalized and saved as a weight file. Typically, when you launch a training process, the model learns from the training data in that epoch and updates the weights and biases, which are saved in a directory you specify before beginning the training process.

---

[1] Two other common and useful functions not covered here are LearningRateScheduler (*https://oreil.ly/CyuGs*) and CSVLogger (*https://oreil.ly/vmeaY*).

However, sometimes you'll want to save the model only if it has improved from the previous epoch, so that the last saved model is always the best model. To do this, you can use the ModelCheck point class. In this section, you are going to see how to leverage this class in your model training process.

Let's try it out using the CIFAR-10 image classification dataset that we used in Chapter 6. As usual, we start by importing the necessary libraries, and then we read the CIFAR-10 data:

```
import tensorflow as tf
from tensorflow.keras import datasets, layers, models
import numpy as np
import matplotlib.pylab as plt
import os
from datetime import datetime

(train_images, train_labels), (test_images, test_labels) =
datasets.cifar10.load_data()
```

First, normalize the pixel values in the images to be in a range of 0 to 1:

```
train_images, test_images = train_images / 255.0,
test_images / 255.0
```

The image labels in this dataset consist of integers. Verify this using a NumPy command:

```
np.unique(train_labels)
```

This shows the values to be:

```
array([0, 1, 2, 3, 4, 5, 6, 7, 8, 9], dtype=uint8)
```

Now you can map these integers to the plain-text labels. The labels here (*https://oreil.ly/hYg5R*) (provided by Alex Krizhevsky, Vinod Nair, and Geoffrey Hinton) are in alphabetical order. Thus, airplane maps to a value of 0 in train_labels, while truck maps to 9:

```
CLASS_NAMES = ['airplane', 'automobile', 'bird', 'cat',
               'deer', 'dog', 'frog', 'horse', 'ship', 'truck']
```

Since there is a separate partition for `test_images`, extract the first 500 images from `test_images` to use for cross validation and name it `validation_images`. You'll use the remaining images for testing.

To use your compute resources more efficiently, convert the `test_images` images and labels from their native NumPy array format to dataset format:

```
validation_dataset = tf.data.Dataset.from_tensor_slices(
(test_images[:500],
 test_labels[:500]))

test_dataset = tf.data.Dataset.from_tensor_slices(
(test_images[500:],
 test_labels[500:]))

train_dataset = tf.data.Dataset.from_tensor_slices(
(train_images,
 train_labels))
```

After executing these commands, you should have all of the images in three datasets: a training dataset (`train_dataset`), a validation dataset (`validation_dataset`), and a testing dataset (`test_dataset`).

It would be nice to know the sizes of these datasets. To find the sample size of a TensorFlow dataset, convert it to a list, and then find the length of the list using the `len` function:

```
train_dataset_size = len(list(train_dataset.as_numpy_iterator()))
print('Training data sample size: ', train_dataset_size)

validation_dataset_size = len(list(validation_dataset.
as_numpy_iterator()))
print('Validation data sample size: ',
validation_dataset_size)

test_dataset_size = len(list(test_dataset.as_numpy_iterator()))
print('Test data sample size: ', test_dataset_size)
```

You can expect these results:

```
Training data sample size:  50000
Validation data sample size:  500
Test data sample size:  9500
```

Next, shuffle and batch the three datasets:

```
TRAIN_BATCH_SIZE = 128
train_dataset = train_dataset.shuffle(50000).batch(
TRAIN_BATCH_SIZE,
drop_remainder=True)

validation_dataset = validation_dataset.batch(
 validation_dataset_size)
test_dataset = test_dataset.batch(test_dataset_size)
```

Notice that `train_dataset` will be split into multiple batches. Each batch will contain `TRAIN_BATCH_SIZE` samples (in this case, 128). Each training batch is fed to the model during the training process to enable incremental updates to weights and biases. There is no need to create multiple batches for validation and testing. These will be used as one batch, but only for the purposes of logging metrics and testing.

Next, specify how often to update weights and validate:

```
STEPS_PER_EPOCH = train_dataset_size // TRAIN_BATCH_SIZE
VALIDATION_STEPS = 1
```

The preceding code means that after the model has seen the number of batches of training data specified by `STEPS_PER_EPOCH`, it's time to test the model with the validation dataset (used as one batch).

To do this, you'll first define the model architecture:

```
model = tf.keras.Sequential([
    tf.keras.layers.Conv2D(32, kernel_size=(3, 3),
      activation='relu',
      kernel_initializer='glorot_uniform', padding='same',
      input_shape = (32,32,3)),
    tf.keras.layers.MaxPooling2D(pool_size=(2, 2)),
    tf.keras.layers.Conv2D(64, kernel_size=(3, 3),
     activation='relu',
      kernel_initializer='glorot_uniform', padding='same'),
    tf.keras.layers.MaxPooling2D(pool_size=(2, 2)),
    tf.keras.layers.Flatten(),
    tf.keras.layers.Dense(256,
     activation='relu', kernel_initializer='glorot_uniform'),
    tf.keras.layers.Dense(10, activation='softmax',
    name = 'custom_class')
```

```
])
model.build([None, 32, 32, 3])
```

Now, compile the model to make sure it's set up properly:

```
model.compile(
        loss=tf.keras.losses.SparseCategoricalCrossentropy(
            from_logits=True),
        optimizer='adam',
        metrics=['accuracy'])
```

Next, name the folders to which TensorFlow should save the model at each checkpoint. Usually, you will rerun the training routine multiple times, and you may find it tedious to create a unique folder name each time. A simple and frequently used approach is to append a timestamp to the model name:

```
MODEL_NAME = 'myCIFAR10-{}'.format(datetime.now().strftime(
"%Y%m%d-%H%M%S"))
print(MODEL_NAME)
```

The preceding command yields a name such as *myCIFAR10-20210123-212138*. You can use this name for the checkpoint directory:

```
checkpoint_dir = './' + MODEL_NAME
checkpoint_prefix = os.path.join(checkpoint_dir, "ckpt-{epoch}")
print(checkpoint_prefix)
```

The preceding command specifies the directory path to be *./ myCIFAR10-20210123-212138/ckpt-{epoch}*. This directory is located one level below your current directory. *{epoch}* will be encoded with the epoch number during training. Now define the myCheckPoint object:

```
myCheckPoint = tf.keras.callbacks.ModelCheckpoint(
    filepath=checkpoint_prefix,
    monitor='val_accuracy',
    mode='max')
```

Here you specified the file path where TensorFlow will save the model at each epoch. You also set it up to monitor validation accuracy.

When you launch the training process with a callback, the callback will expect a Python list. So, let's put the myCheckPoint object into a Python list:

```
myCallbacks = [
    myCheckPoint
]
```

Now launch the training process. This command assigns the entire model training history to the object hist, which is a Python dictionary:

```
hist = model.fit(
    train_dataset,
    epochs=12,
    steps_per_epoch=STEPS_PER_EPOCH,
    validation_data=validation_dataset,
    validation_steps=VALIDATION_STEPS,
    callbacks=myCallbacks).history
```

You can view the cross-validation accuracy from the first epoch of training to the last using the command hist['val_accuracy']. The display should look something like this:

```
[0.47200000286102295,
 0.5680000185966492,
 0.6000000238418579,
 0.5899999737739563,
 0.6119999885559082,
 0.6019999980926514,
 0.6100000143051147,
 0.6380000114440918,
 0.6100000143051147,
 0.5699999928474426,
 0.5619999766349792,
 0.5960000157356262]
```

In this case, cross-validation accuracy improved for a number of epochs, then gradually degraded. This degradation is a typical sign of overfitting. The best model here is the one with the highest validation accuracy (the highest value in the array). To determine its position (or index) in the array, use this code:

```
max_value = max(hist['val_accuracy'])
max_index = hist['val_accuracy'].index(max_value)
print('Best epoch: ', max_index + 1)
```

Remember to add 1 to `max_index` because the epoch starts at 1, not 0 (unlike a NumPy array index). The output is:

```
Best epoch:  8
```

Next, take a look at the checkpoint directories by running the following Linux command in your Jupyter Notebook cell:

```
!ls -lrt ./cifar10_training_checkpoints
```

You will see the contents of this directory (shown in Figure 7-1).

```
total 48
drwxr-xr-x 4 root root 4096 Jan 14 01:56 ckpt_1
drwxr-xr-x 4 root root 4096 Jan 14 01:56 ckpt_2
drwxr-xr-x 4 root root 4096 Jan 14 01:56 ckpt_3
drwxr-xr-x 4 root root 4096 Jan 14 01:57 ckpt_4
drwxr-xr-x 4 root root 4096 Jan 14 01:57 ckpt_5
drwxr-xr-x 4 root root 4096 Jan 14 01:57 ckpt_6
drwxr-xr-x 4 root root 4096 Jan 14 01:57 ckpt_7
drwxr-xr-x 4 root root 4096 Jan 14 01:57 ckpt_8
drwxr-xr-x 4 root root 4096 Jan 14 01:57 ckpt_9
drwxr-xr-x 4 root root 4096 Jan 14 01:57 ckpt_10
drwxr-xr-x 4 root root 4096 Jan 14 01:57 ckpt_11
drwxr-xr-x 4 root root 4096 Jan 14 01:57 ckpt_12
```

Figure 7-1. Model saved at each checkpoint

You can rerun this command and specify a specific directory to see the model built by a particular epoch (as shown in Figure 7-2):

```
!ls -lrt ./cifar10_training_checkpoints/ckpt_8
```

```
total 136
drwxr-xr-x 2 root root   4096 Jan 14 01:57 variables
drwxr-xr-x 2 root root   4096 Jan 14 01:57 assets
-rw-r--r-- 1 root root 127273 Jan 14 01:57 saved_model.pb
```

Figure 7-2. Model files saved at checkpoint 8

So far, you have seen how to use `CheckPoint` to save the model at each epoch. If you wish to save only the best model, specify `save_best_only = True`:

```
best_only_checkpoint_dir =
 './best_only_cifar10_training_checkpoints'
best_only_checkpoint_prefix = os.path.join(
best_only_checkpoint_dir,
"ckpt_{epoch}")

bestCheckPoint = tf.keras.callbacks.ModelCheckpoint(
    filepath=best_only_checkpoint_prefix,
    monitor='val_accuracy',
    mode='max',
    save_best_only=True)
```

Then put bestCheckPoint in a callback list:

```
    bestCallbacks = [
    bestCheckPoint
]
```

After that, you can launch the training process:

```
best_hist = model.fit(
    train_dataset,
    epochs=12,
    steps_per_epoch=STEPS_PER_EPOCH,
    validation_data=validation_dataset,
    validation_steps=VALIDATION_STEPS,
    callbacks=bestCallbacks).history
```

In this training, rather than saving all checkpoints, bestCall
backs causes the model to be saved only if it has a better validation accuracy than the previous epoch. The save_best_only
option lets you save checkpoints after the first epoch *only* if
there is an incremental improvement to the model metric of
your choice (specified with monitor), so that the last checkpoint
saved is the best model.

To look at what you've saved, run the following command in a
Jupyter Notebook cell:

```
!ls -lrt ./best_only_cifar10_training_checkpoints
```

The saved models with incremental improvement in validation
accuracy are shown in Figure 7-3.

```
total 12
drwxr-xr-x 4 root root 4096 Jan 14 01:57 ckpt_1
drwxr-xr-x 4 root root 4096 Jan 14 01:58 ckpt_3
drwxr-xr-x 4 root root 4096 Jan 14 01:58 ckpt_9
```

Figure 7-3. Models saved with save_best_only set to True

The model from the first epoch is always saved. In the third epoch, the model shows improvement in validation accuracy, so the third checkpoint model is saved. The training continues. Validation accuracy improves in the ninth epoch, so the ninth checkpoint model is the last directory that is saved. The training continues through the 12th epoch without any further incremental improvement in validation accuracy. This means that the ninth checkpoint directory contains the model with the best validation accuracy.

Now that you're familiar with ModelCheckpoint, let's examine another callback object: EarlyStopping.

EarlyStopping

The EarlyStopping callback object enables you to stop the training process before it reaches the final epoch. Usually, you'd do this to save training time if the model isn't improving.

The object lets you specify a model metric—for example, validation accuracy—to monitor through all epochs. If the specified metric does not improve after a certain number of epochs, the training will stop.

To define an EarlyStopping object, use this command:

```
myEarlyStop = tf.keras.callbacks.EarlyStopping(
monitor='val_accuracy',
patience=4)
```

In this case, you are monitoring validation accuracy at each epoch. You set the patience parameter to 4, which means that if validation accuracy does not improve within four epochs, the training stops.

TIP

To learn more ways to customize early stopping, see the TensorFlow 2 documentation (*https://oreil.ly/UDpaA*).

To implement early stopping with a `ModelCheckpoint` object in a callback, you need to put it into a list:

```
myCallbacks = [
    myCheckPoint,
    myEarlyStop
]
```

The training process is the same, but you designate `call backs=myCallbacks`:

```
hist = model.fit(
    train_dataset,
    epochs=20,
    steps_per_epoch=STEPS_PER_EPOCH,
    validation_data=validation_dataset,
    validation_steps=VALIDATION_STEPS,
    callbacks=myCallbacks).history
```

Once you launch the preceding training command, the output should resemble Figure 7-4.

```
Epoch 13/20
250/250 [==============================] - 4s 14ms/step - loss: 0.3992 - accuracy: 0.8647 - val_loss: 0.9272 - val_accuracy: 0.7120
INFO:tensorflow:Assets written to: ./myCIFAR10-20210123-212138/ckpt-13/assets
Epoch 14/20
250/250 [==============================] - 4s 14ms/step - loss: 0.3373 - accuracy: 0.8889 - val_loss: 1.0225 - val_accuracy: 0.6800
INFO:tensorflow:Assets written to: ./myCIFAR10-20210123-212138/ckpt-14/assets
Epoch 15/20
250/250 [==============================] - 4s 14ms/step - loss: 0.2986 - accuracy: 0.9027 - val_loss: 1.0352 - val_accuracy: 0.7220
INFO:tensorflow:Assets written to: ./myCIFAR10-20210123-212138/ckpt-15/assets
Epoch 16/20
250/250 [==============================] - 4s 14ms/step - loss: 0.2534 - accuracy: 0.9215 - val_loss: 1.0596 - val_accuracy: 0.7200
INFO:tensorflow:Assets written to: ./myCIFAR10-20210123-212138/ckpt-16/assets
Epoch 17/20
250/250 [==============================] - 4s 13ms/step - loss: 0.2129 - accuracy: 0.9337 - val_loss: 1.1308 - val_accuracy: 0.7160
INFO:tensorflow:Assets written to: ./myCIFAR10-20210123-212138/ckpt-17/assets
Epoch 18/20
250/250 [==============================] - 4s 13ms/step - loss: 0.1820 - accuracy: 0.9435 - val_loss: 1.1767 - val_accuracy: 0.6960
INFO:tensorflow:Assets written to: ./myCIFAR10-20210123-212138/ckpt-18/assets
Epoch 19/20
250/250 [==============================] - 4s 14ms/step - loss: 0.1449 - accuracy: 0.9574 - val_loss: 1.2077 - val_accuracy: 0.7100
INFO:tensorflow:Assets written to: ./myCIFAR10-20210123-212138/ckpt-19/assets
```

Figure 7-4. Early stop during training

In the training shown in Figure 7-4, the best validation accuracy appears in epoch 15, with a value of 0.7220. After four more epochs, validation accuracy has not improved beyond that value, and so the training stops after epoch 19.

Summary

While the `ModelCheckpoint` class lets you set a condition or cadence to save the model during training, the `EarlyStopping` class lets you terminate training early if the model is showing no improvement to the metric of your choice. Together, these classes are specified in a Python list, and this list is passed into the training routine as a callback.

Many other functions are available for monitoring training progress (see tf.keras.callbacks.Callback (*https://oreil.ly/1nIE6*) and the Keras Callbacks API (*https://oreil.ly/BeJBW*)), but `ModelCheckpoint` and `EarlyStopping` are two of the most frequently used ones.

The remainder of this chapter will dive deep into the popular callback class known as `TensorBoard`, which provides a visual representation of your training progress and results.

TensorBoard

If you wish to visualize your model and training process, TensorBoard is the tool for you. TensorBoard provides a visual representation of how your model parameters and metrics evolve from the beginning to the end of your training process. It's frequently used for tracking model accuracy over training epochs. It can also let you see how weights and biases evolve in each model layer. And just like `ModelCheckpoint` and `EarlyStopping`, TensorBoard is applied to the training process via the callbacks module. You create an object that represents a `TensorBoard`, then pass that object as a member of a callback list.

Let's try building a model that classifies CIFAR-10 images. As usual, start by importing libraries, loading the CIFAR-10 images, and normalizing pixel values to a range of 0 to 1:

```
import tensorflow as tf
from tensorflow.keras import datasets, layers, models
import numpy as np
import matplotlib.pylab as plt
import os
```

```
from datetime import datetime

(train_images, train_labels), (test_images, test_labels) =
datasets.cifar10.load_data()

train_images, test_images = train_images / 255.0,
 test_images / 255.0
```

Define your plain-text labels:

```
CLASS_NAMES = ['airplane', 'automobile', 'bird', 'cat',
               'deer','dog', 'frog', 'horse', 'ship', 'truck']
```

Now convert the images to a dataset:

```
validation_dataset = tf.data.Dataset.from_tensor_slices(
(test_images[:500], test_labels[:500]))

test_dataset = tf.data.Dataset.from_tensor_slices(
(test_images[500:], test_labels[500:]))

train_dataset = tf.data.Dataset.from_tensor_slices(
(train_images, train_labels))
```

Then determine the data size for training, validation, and test
partitions:

```
train_dataset_size = len(list(train_dataset.as_numpy_iterator()))
print('Training data sample size: ', train_dataset_size)

validation_dataset_size = len(list(validation_dataset.
as_numpy_iterator()))
print('Validation data sample size: ',
 validation_dataset_size)

test_dataset_size = len(list(test_dataset.as_numpy_iterator()))
print('Test data sample size: ', test_dataset_size)
```

Your results should look like this:

```
Training data sample size:  50000
Validation data sample size:  500
Test data sample size:  9500
```

Now you can shuffle and batch the data:

```
TRAIN_BATCH_SIZE = 128
train_dataset = train_dataset.shuffle(50000).batch(
TRAIN_BATCH_SIZE,
drop_remainder=True)
```

```
validation_dataset = validation_dataset.batch(
validation_dataset_size)
test_dataset = test_dataset.batch(test_dataset_size)
```

And then specify parameters for setting up a cadence to update model weights:

```
STEPS_PER_EPOCH = train_dataset_size // TRAIN_BATCH_SIZE
VALIDATION_STEPS = 1
```

STEPS_PER_EPOCH is an integer, rounded down from the division between train_dataset_size and TRAIN_BATCH_SIZE. (The double forward slashes indicate division and rounding down to the nearest integer.)

We'll reuse the model architecture we built in "ModelCheckpoint" on page 130:

```
model = tf.keras.Sequential([
    tf.keras.layers.Conv2D(32,
      kernel_size=(3, 3),
      activation='relu',
      name = 'conv_1',
      kernel_initializer='glorot_uniform',
      padding='same', input_shape = (32,32,3)),
    tf.keras.layers.MaxPooling2D(pool_size=(2, 2)),
    tf.keras.layers.Conv2D(64, kernel_size=(3, 3),
      activation='relu', name = 'conv_2',
        kernel_initializer='glorot_uniform', padding='same'),
    tf.keras.layers.MaxPooling2D(pool_size=(2, 2)),
    tf.keras.layers.Conv2D(64,
      kernel_size=(3, 3),
      activation='relu',
      name = 'conv_3',
        kernel_initializer='glorot_uniform', padding='same'),
    tf.keras.layers.Flatten(name = 'flat_1',),
    tf.keras.layers.Dense(64, activation='relu',
      kernel_initializer='glorot_uniform',
      name = 'dense_64'),
    tf.keras.layers.Dense(10,
      activation='softmax',
      name = 'custom_class')
])
model.build([None, 32, 32, 3])
```

Notice that this time, each layer has a name. Designating a name for each layer helps you know which layer you are inspecting. This is not required, but it's a good practice for visualization in TensorBoard.

Now compile the model to make sure the model architecture is valid, and designate a loss function:

```
model.compile(
        loss=tf.keras.losses.SparseCategoricalCrossentropy(
            from_logits=True),
        optimizer='adam',
        metrics=['accuracy'])
```

Setting a model name will be helpful later, when TensorBoard lets you pick a model (or models) and inspect the visualization of its training results. You can append a timestamp to the model name as we did when using ModelCheckpoint:

```
MODEL_NAME =
'myCIFAR10-{}'.format(datetime.now().strftime("%Y%m%d-%H%M%S"))

print(MODEL_NAME)
```

In this example, MODEL_NAME is myCIFAR10-20210124-135804. Yours will be similar.

Next, set up the checkpoint directory:

```
checkpoint_dir = './' + MODEL_NAME
checkpoint_prefix = os.path.join(checkpoint_dir, "ckpt-{epoch}")
print(checkpoint_prefix)
```

./myCIFAR10-20210124-135804/ckpt-{epoch} is the name of this checkpoint directory.

Define the model checkpoint:

```
myCheckPoint = tf.keras.callbacks.ModelCheckpoint(
    filepath=checkpoint_prefix,
    monitor='val_accuracy',
mode='max')
```

Next you will define a TensorBoard, and then we'll take a closer look at this code:

```
myTensorBoard = tf.keras.callbacks.TensorBoard(
log_dir='./tensorboardlogs/{}'.format(MODEL_NAME),
write_graph=True,
write_images=True,
histogram_freq=1)
```

The first argument specified here is `log_dir`. It is the path where you want to save the training logs. As indicated, it is in a directory below the current level, named *tensorboardlogs*, which is followed by a subdirectory named `MODEL_NAME`. As training progresses, your logs will be generated and stored here so that TensorBoard can parse them for visualization.

The parameter `write_graph` is set to True, so that the model graph will be visualized. Another parameter, `write_images`, is also set to True. This ensures that model weights will be written in the log, so you can visualize how they change throughout training.

Finally, `histogram_freq` is set to 1. This tells TensorBoard when to create a visualization by epoch: 1 means a visualization is created for each epoch. For more parameters, see the Tensor-Board documentation (*https://oreil.ly/k1Pd2*).

Finally, you have two callback objects to set up: `myCheckPoint` and `myTensorBoard`. To put both in a Python list, you only have to do this:

```
myCallbacks = [
    myCheckPoint,
    myTensorBoard
]
```

Then pass your `myCallbacks` list into the training routine:

```
hist = model.fit(
    train_dataset,
    epochs=30,
    steps_per_epoch=STEPS_PER_EPOCH,
    validation_data=validation_dataset,
    validation_steps=VALIDATION_STEPS,
    callbacks=myCallbacks).history
```

Once the training process is done, there are three ways to invoke TensorBoard. You can run it from the next cell in your Jupyter Notebook on your own computer, from a command terminal on your own computer, or in Google Colab. We'll look at each of these options in turn.

Invoking TensorBoard by Local Jupyter Notebook

If you choose to use your Jupyter Notebook, run the following command in the next cell:

```
!tensorboard --logdir='./tensorboardlogs/'
```

Notice that in this case, when you specify the path to find the training logs, the argument is `logdir`, not `log_dir` as it was when you defined `myTensorBoard`.

Once you run the preceding command, you'll see this:

```
Serving TensorBoard on localhost; to expose to the network, use a
proxy or pass --bind_all
TensorBoard 2.3.0 at http://localhost:6006/ (Press CTRL+C to quit)
```

As you can see, TensorBoard is running at your current compute instance (localhost) on port number 6006.

Now open a browser and navigate to *http://localhost:6006*, and you will see TensorBoard running, as shown in Figure 7-5.

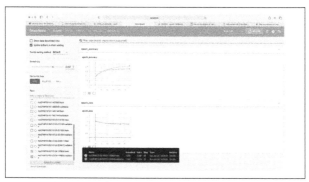

Figure 7-5. TensorBoard visualization

As you can see, through each epoch, accuracy and loss are traced in graphs. Training data is shown in lighter gray and validation data is shown in a darker gray.

The main advantage of using a Jupyter Notebook cell is convenience. A drawback is that the cell that runs the `!tensorboard` command will remain active, and you won't be able to use this notebook until you stop TensorBoard.

Invoking TensorBoard by Local Command Terminal

Your second option is to launch TensorBoard from a command terminal in your local environment. As shown in Figure 7-6, the equivalent command is:

```
tensorboard --logdir='./tensorboardlogs/'
```

```
[tf23] (base) mbp16@casablancas-MacBook-Pro chapter_07 % tensorboard --logdir='./tensorboardlogs/'
Serving TensorBoard on localhost; to expose to the network, use a proxy or pass --bind_all
TensorBoard 2.3.0 at http://localhost:6006/ (Press CTRL+C to quit)
```

Figure 7-6. TensorBoard invoked from the command terminal

Remember that logdir is the directory path to the training logs created by the training callback API. The command in the preceding code uses relative path notation; you may use a full path if you prefer.

The output is exactly the same as seen in the Jupyter Notebook: the URL (*http://localhost:6006*). Open this URL with a browser to display TensorBoard.

Invoking TensorBoard by Colab Notebook

Now for our third option. If you are using a Google Colab notebook for this exercise, then invoking TensorBoard will be a little different from what you have seen so far. You won't be able to open a browser on your local computer to point to the Colab notebook, because it will be running in Google's cloud environment. You'll thus need to install a TensorBoard notebook extension. This may be done in the first cell, when you import all the libraries. Just add this command and run it in the first Colab cell:

```
%load_ext tensorboard
```

Once this is done, whenever you are ready to invoke TensorBoard (such as after training is complete), use this command:

```
%tensorboard --logdir ./tensorboardlogs/
```

You will see the output running inside your Colab notebook, looking as shown in Figure 7-5.

Visualizing Model Overfitting Using TensorBoard

When you use TensorBoard as a callback for model training, you will get graphs of model accuracy and loss from the first epoch to the last.

For example, with our CIFAR-10 image classification model, you'll see output like that shown in Figure 7-5. In that particular training run, while both training and validation accuracy are increasing and losses are decreasing, both trends start to flatten out, indicating that further training epochs are likely to deliver only marginal gains.

Note also that in this run, validation accuracy is lower than training accuracy, while validation loss is higher than training loss. This makes sense, because the model performs better with training data than when tested with cross-validation data.

You may also get graphs in the Scalars tab of TensorBoard, like those shown in Figure 7-7.

In Figure 7-7, the darker lines indicate validation metrics, while lighter gray lines indicate training metrics. Model accuracy in the training data is much higher than in the cross-validation data, and the loss in the training data is much lower than in the cross-validation data.

You may also notice that cross-validation accuracy peaked at epoch 10, with a value slightly above 0.7. After that, validation data accuracy started to decrease, while loss started to increase. This is a clear sign of model overfitting. What these graphs are telling you is that after epoch 10, your model started to memorize training data patterns. That doesn't help when it encounters new, previously unseen data (like the cross-validation images). In fact, the model's performance in cross validation (accuracy and loss) will start to worsen.

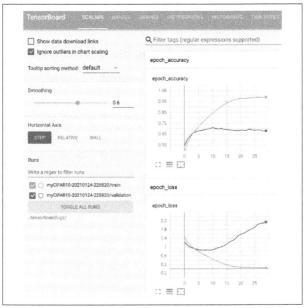

Figure 7-7. Model overfitting shown in TensorBoard

Once you inspect these graphs, you'll know which epoch delivered the best model of the training process. You'll also be well informed about when the model starts to overfit and memorize its training data.

If your model still has room for improvement, like the one in Figure 7-5, you may decide to increase your training epochs and keep looking for the best model before the overfitting pattern starts to appear (see Figure 7-7).

Visualizing the Learning Process Using TensorBoard

Another cool feature in TensorBoard is the histogram of weights and bias distribution. These are shown across each epoch as the result of training. By visualizing how these parameters are distributed and how their distribution changes over time, you gain insights into the impact of the training process.

Let's look at how to use TensorBoard to inspect model weights and bias distributions. This information will be in Tensor-Board's Histogram tab (Figure 7-8).

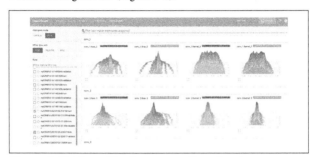

Figure 7-8. Weights and bias histogram in TensorBoard

On the left is the panel with all the models trained. Notice there are two models selected. On the right are their weights (denoted as `kernel_0`) and bias distributions across each epoch of training. Each row of figures represents a particular layer in the model. The first layer is named `conv_1`, which is what you named this layer back when you set up the model architecture.

Let's examine these figures a bit more closely. We'll start with the conv_1 layer shown in Figure 7-9.

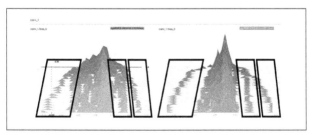

Figure 7-9. Bias distribution in conv_1 layer through training

In both models, the distribution of bias values in the conv_1 layer definitely changed from the first epoch (background) to the last epoch (foreground). The boxes indicate that as training progresses, a certain distribution pattern of bias starts to emerge in all the nodes of this layer. The new values are away from zero, or the center of the overall distribution.

Let's also take a look at the distribution of weights. This time, let's just focus on one model and one layer: conv_3. This is shown in Figure 7-10.

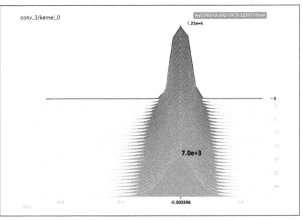

Figure 7-10. Weight distribution in conv_3 layer through training

What's worth noting is that the distribution became broader and flatter as training progressed. This can be seen from the peak count of the histogram from the first to the last epochs, from 1.22e+4 to 7.0e+3. This means that the histogram gradually becomes broader, and that there are more weights being updated with values away from zero (the center of the histogram).

Using TensorBoard, you can examine different layers and combinations of model training runs to see how they are impacted by the training process or by changes in model architecture.

This is why TensorBoard is frequently used to visually inspect the model training process.

Wrapping Up

In this chapter, you saw some of the most popular methods for tracking the model training process. This chapter presented the concept of model checkpointing and offered two important ways to help you manage how to save models during training: either save the model at every epoch, or only save the model when an incremental improvement in model metrics is validated. You also learned that model accuracy in cross validation determines when the model starts to overfit the training data.

Another important tool you learned about in this chapter is TensorBoard, which can be used to visualize the training process. TensorBoard presents visual images of basic metrics (accuracy and loss) trends through training epochs. It also lets you inspect the weight and bias distributions of each layer throughout the training. All these techniques are easy to implement in the training routine via callbacks.

In the next chapter, you will see how to implement distributed training in TensorFlow, which takes advantage of performant compute units such as GPUs to provide shorter training times.

## CHAPTER 8

Distributed Training

Training a machine learning model may take a long time, especially if your training dataset is huge or you are using a single machine to do the training. Even if you have a GPU card at your disposal, it can still take weeks to train a complex model such as ResNet50, a computer vision model with 50 convolution layers, trained to classify objects into a thousand categories.

Reducing model training time requires a different approach. You already saw some of the options available: in Chapter 5, for example, you learned to leverage datasets in a data pipeline. Then there are more powerful accelerators, such as GPUs and TPUs (which are exclusively available in Google Cloud).

This chapter will cover a different way to train your model, known as *distributed training*. Distributed training runs a model training process in parallel on a cluster of devices, such as CPUs, GPUs, and TPUs, to speed up the training process. (In this chapter, for the sake of concision, I will refer to hardware accelerators such as GPUs, CPUs, and TPUs as *workers* or *devices*.) After you read this chapter, you will know how to refactor your single-node training routine for distributed training. (Every example you have seen in this book up to this point has been single node: that is, they have all used a machine with one CPU to train the model.)

In distributed training, your model is trained by multiple independent processes. You can think of each process as an independent training endeavor. Each process runs the training routine on a separate device, using a subset (called a *shard*) of the training data. This means that each process uses different training data. As each process completes an epoch of training, it sends the results back to a *master routine*, which collects and aggregates the results and then issues updates to all of the processes. Each process then resumes training with the updated weights and biases.

Before we dive into the implementation code, let's take a closer look at the heart of distributed ML model training. We will start with the concept of data parallelism.

Data Parallelism

The first thing you need to understand about distributed training is how training data is handled. The predominant architecture in distributed training is known as *data parallelism*. In this architecture, you run the same model and computation logic on each worker. Each worker computes the loss and gradients using a shard of data that is different from those of the other workers, then uses these gradients to update the model parameters. The updated model in each individual worker is then used in the next round of computation. This concept is illustrated in Figure 8-1.

Two common approaches are designed to update the model with these gradients: asynchronous parameter servers and synchronous allreduce. We'll look at each in turn.

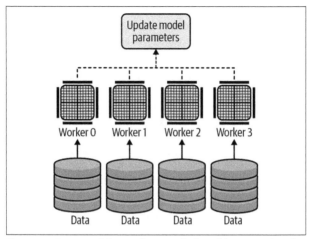

Figure 8-1. Data parallelism architecture (Adapted from Distributed TensorFlow training (https://oreil.ly/beSob) in Google I/O 2018 video)

Asynchronous Parameter Server

Let's look first at the *asynchronous parameter server* approach, shown in Figure 8-2.

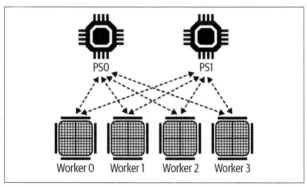

Figure 8-2. Distributed training using asynchronous parameter servers. (Adapted from Distributed TensorFlow training (https://oreil.ly/beSob) in Google I/O 2018 video)

The devices labeled PS0 and PS1 in Figure 8-2 are *parameter servers*; these servers hold the parameters of your model. Other devices are designated as workers, as labeled in Figure 8-2.

The workers do the bulk of the computation. Each worker fetches the parameters from the server, computes loss and gradients, and then sends the gradients back to the parameter server, which uses them to update the model's parameters. Each worker does this independently, so this approach can be scaled up to use a large number of workers. The advantage here is that if training workers are preempted by high-priority production jobs, if there is asymmetry between the workers, or if a machine goes down for maintenance, it doesn't hurt your scaling, because the workers are not waiting on each other.

However, there is a downside: the workers can get out of sync. This can lead to computing gradients on stale parameter values, which can delay model convergence and therefore delay training toward the best model. With the recent popularity and prevalence of hardware accelerators, this approach is implemented less frequently than synchronous allreduce, which we'll discuss next.

Synchronous Allreduce

The *synchronous allreduce* approach has become more common as fast hardware accelerators such as GPUs and TPUs have become more widely available.

In a synchronous allreduce architecture (shown in Figure 8-3), each worker holds a copy of the model's parameters in its own memory. There are no parameter servers. Instead, each worker computes the loss and gradients based on a shard of training samples. Once that computation is complete, the workers communicate among themselves to propagate the gradients and update the model parameters. All workers are synchronized, which means the next round of computation begins *only* when each worker has received the new gradients and updated the model parameters in its memory accordingly.

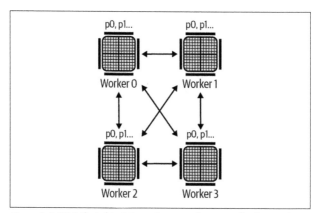

Figure 8-3. Distributed training using a synchronous allreduce architecture (adapted from Distributed TensorFlow training (https://oreil.ly/beSob) in Google I/O 2018 video)

With these fast devices in a connected cluster, differences in processing time between workers are not an issue. As a result, this approach usually leads to a faster convergence on the best model than the asynchronous parameter server architecture.

Allreduce is a type of algorithm that combines gradients across different workers. This algorithm aggregates gradient values from different workers by, for example, summing them and then copying them to different workers. Its implementation can be very efficient as it reduces the overhead involved in synchronizing the gradients. Many allreduce algorithm implementations are available, depending on the types of communication available between workers and on the architecture's topology. A common implementation of this algorithm, known as *ring-allreduce*, is shown in Figure 8-4.

In a ring-allreduce implementation, each worker sends its gradient to its successor on the ring and receives a gradient from its predecessor. Eventually, each worker receives a copy of the combined gradients. Ring-allreduce uses network bandwidth optimally, because it uses both the upload and download bandwidth of each worker. It's fast whether working with multiple workers on a single machine or a small number of machines.

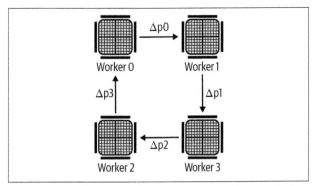

Figure 8-4. Ring-allreduce implementation (Adapted from Distributed TensorFlow training (https://oreil.ly/beSob) in Google I/O 2018 video)

Now let's see how you can do all this in TensorFlow. We're going to focus on scaling to multiple GPUs with a synchronous allreduce architecture. You'll see how easy it is to refactor your single-node training code for allreduce. That's because these high-level APIs handle a lot of the aforementioned complexity and nuances of data parallelism for you, behind the scenes.

Using the Class tf.distribute.MirroredStrategy

The easiest way to implement distributed training is to use the `tf.distribute.MirroredStrategy` class provided by Tensor-Flow. (For the details of various distributed training strategies supported by TensorFlow, see "Types of strategies" in the TensorFlow documentation (*https://oreil.ly/0jQed*)). As you will see, implementing this class requires only minimal changes in your source code, and you will still be able to run in single-node mode, so you don't have to worry about backward compatibility. It also takes care of updating weights and biases, metrics, and model checkpoints for you. Further, you don't have to worry about how to split training data into shards for each device. You don't need to write code to handle retrieving or updating parameters from each device. Nor do you need to worry about how to ensure that gradients and losses across devices are aggregated correctly. The distribution strategy does all of that for you.

We'll look briefly at a few code snippets that demonstrate the changes you need to make in your training code for a case of one machine with multiple devices:

1. Create an object to handle distributed training. You can do this at the beginning of the source code:

```
strategy = tf.distribute.MirroredStrategy()
```

The `strategy` object contains a property that holds the number of devices available in the machine. You may use this command to show how many GPUs or TPUs are at your disposal:

```
print('Number of devices: {}'.format(
 strategy.num_replicas_in_sync))
```

If you are using a GPU cluster, such as through Databricks or a cloud provider's environment, you will see the number of GPUs to which you have access:

```
Number of devices: 2
```

Note that Google Colab only provides one GPU for each user.

2. Wrap your model definition and loss function in a `strategy` scope. You just need to make sure the model definition and compilation, including the loss function of your choice, are encapsulated in a specific scope:

```
with strategy.scope():
  model = tf.keras.Sequential([
  ...
```

```
  ])
  model.build(…)
  model.compile(
    loss=tf.keras.losses…,
    optimizer=…,
    metrics=…)
```

These are the only two places you need to make code changes.

The `tf.distribute.MirroredStrategy` class is the workhorse behind the curtain. As you have seen, the `strategy` object we created knows how many devices are available. This information enables it to split the training data into different shards and feed each shard into a particular device. Since the model architecture is wrapped in this object's scope, it is also held in each device's memory. This allows each device to run the training routine on the same model architecture, minimize the same loss function, and update the gradients according to its specific shard of training data. The model architecture and parameters are replicated, or *mirrored*, across all devices. The `Mirrored Strategy` class also implements the ring-allreduce algorithm behind the scenes, so you don't have to worry about aggregating all the gradients from each device.

The class is aware of your hardware settings and their potential for distributed training, so there is no need for you to change your `model.fit` training routine or data ingestion method. Saving model checkpoints and model summaries works the same way as it does in single-node training, as we saw in "Model-Checkpoint" on page 130.

Setting Up Distributed Training

To try the distributed training examples in this chapter, you will need access to multiple GPUs or TPUs. For simplicity, consider using one of the various commercial platforms that offer GPU clusters, such as Databricks (*https://databricks.com*) and Paperspace (*https://www.paperspace.com*). Other choices include major cloud vendors, which offer a wide variety of platforms, from managed services to containers. For the sake of simplicity and ease of availability, the examples in this chapter are done in Databricks, a cloud-based compute vendor. It allows you to set up a distributed compute cluster of either GPUs or CPUs to run heavy workloads that a single-node machine cannot handle.

While Databricks offers a free "community edition," it does not provide access to GPU clusters; for that you will need to create a paid account (*https://oreil.ly/byE1d*). Then you can associate Databricks with a cloud vendor of your choice and create a GPU cluster with the configuration shown in Figure 8-5. My advice: once you're done with your work, download your note-book and delete the clusters you created.

Figure 8-5. Setting up a Databricks GPU cluster

You may notice in Figure 8-5 that there are Autopilot Options. The Enable autoscaling option will automatically scale to more workers as needed, based on the workload. To save costs, I also set the option for this cluster to terminate after 120 minutes of inactivity. (Note that terminating the cluster does *not* mean you have deleted it. It remains and continues incurring a small charge in your account until you delete it.) Once you complete the configuration, click the Create Cluster button at the top. It usually takes about 10 minutes to complete the process.

Next, create a notebook (Figure 8-6).

Figure 8-6. Creating a notebook in the Databricks environment

Give your notebook a name, ensure the default language is set to Python, and select the cluster that you just created (Figure 8-7). Click the Create button to generate a blank notebook.

Create Notebook

Name	My Distributed Training
Default Language	Python
Cluster	Select a cluster

Cancel   Create

Figure 8-7. Setting up your notebook

Now make sure your notebook is attached to the GPU cluster (Figure 8-8).

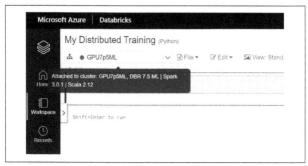

Figure 8-8. Attaching your notebook to an active cluster

Now go ahead and start the GPU cluster. Then go to the Libraries tab and click the Install New button, as shown in Figure 8-9.

Figure 8-9. Installing libraries in a Databricks cluster

When the Install Library prompt appears (Figure 8-10), select PyPI as the Library Source, and in the Package field type **tensorflow-datasets**, and then click the Install button.

Once you have done this, you will be able to use TensorFlow's dataset API to work through the examples in this chapter. In the next section, you are going to see how to use a Databricks notebook to try out distributed training with the GPU cluster you just created.

Figure 8-10. Installing the tensorflow-datasets library in a Databricks cluster

Using a GPU Cluster with tf.distribute.MirroredStrategy

In Chapter 7, you used the CIFAR-10 image dataset to build an image classifier with single-node training. In this example, you will instead train a classifier using the distributed training method.

As usual, the first thing you need to do is import the necessary libraries:

```
import tensorflow as tf
from tensorflow.keras import datasets, layers, models
import numpy as np
import os
from datetime import datetime
```

Now is a good time to create a MirroredStrategy object to handle distributed training:

```
strategy = tf.distribute.MirroredStrategy()
```

Your output should look something like this:

```
INFO:tensorflow:Using MirroredStrategy with devices
('/job:localhost/replica:0/task:0/device:GPU:0',
'/job:localhost/replica:0/task:0/device:GPU:1')
```

This shows that there are two GPUs. You can confirm this with the following statement, as we did previously:

```
print('Number of devices: {}'.format(
strategy.num_replicas_in_sync))
Number of devices: 2
```

Now load the training data and normalize each image pixel range to be between 0 and 1:

```
(train_images, train_labels), (test_images, test_labels) =
datasets.cifar10.load_data()
# Normalize pixel values to be between 0 and 1
train_images, test_images = train_images / 255.0,
test_images / 255.0
```

You can define plain-text labels in a list:

```
# Plain-text name in alphabetical order. See
https://www.cs.toronto.edu/~kriz/cifar.html
CLASS_NAMES = ['airplane', 'automobile', 'bird', 'cat',
              'deer','dog', 'frog', 'horse', 'ship', 'truck']
```

These plain-text labels, from the CIFAR-10 dataset (*https://oreil.ly/fCvCX*), are in alphabetical order: "airplane" maps to the value 0 in `train_labels`, while "truck" maps to 9.

Since there is a separate partition for `test_images`, extract the first 500 images in `test_images` to use as validation images, while keeping the remainder for testing. Also, to more efficiently using compute resources, convert these images and labels from their native NumPy array format to the dataset format:

```
validation_dataset = tf.data.Dataset.from_tensor_slices(
(test_images[:500],
 test_labels[:500]))

test_dataset = tf.data.Dataset.from_tensor_slices(
(test_images[500:],
 test_labels[500:]))

train_dataset = tf.data.Dataset.from_tensor_slices(
(train_images,
 train_labels))
```

After you execute these commands, you'll have all of the images in the formats of training dataset, validation dataset, and test dataset.

It would be nice to know your dataset's size. To find out the sample size of a TensorFlow dataset, convert it to a list, then find the length of the list using the `len` function:

```
train_dataset_size = len(list(train_dataset.as_numpy_iterator()))
print('Training data sample size: ', train_dataset_size)

validation_dataset_size = len(list(validation_dataset.
as_numpy_iterator()))
print('Validation data sample size: ', validation_dataset_size)

test_dataset_size = len(list(test_dataset.as_numpy_iterator()))
print('Test data sample size: ', test_dataset_size)
```

You can expect these results:

```
Training data sample size:  50000
Validation data sample size:  500
Test data sample size:  9500
```

Next, shuffle and batch the three datasets:

```
TRAIN_BATCH_SIZE = 128
train_dataset = train_dataset.shuffle(50000).batch(
TRAIN_BATCH_SIZE, drop_remainder=True)

validation_dataset = validation_dataset.batch(validation_dataset_size)
test_dataset = test_dataset.batch(test_dataset_size)
```

Notice that `train_dataset` will be split into multiple batches, where each batch contains `TRAIN_BATCH_SIZE` samples. Each training batch is fed to the model during the training process to enable incremental updates to the weights and biases. There is no need to create multiple batches for validation and testing: these will be used as one batch, for the purposes of metrics logging and testing only.

Next, specify how often weight updates and validation should occur:

```
STEPS_PER_EPOCH = train_dataset_size // TRAIN_BATCH_SIZE
VALIDATION_STEPS = 1
```

The preceding code means that after the model has seen STEPS_PER_EPOCH batches of training data, it's time to test it with the validation dataset, used as one batch.

Now you need to wrap the model definition, model compilation, and loss function inside the strategy scope:

```python
with strategy.scope():
  model = tf.keras.Sequential([
    tf.keras.layers.Conv2D(32, kernel_size=(3, 3),
activation='relu', name = 'conv_1',
      kernel_initializer='glorot_uniform',
padding='same', input_shape = (32,32,3)),
    tf.keras.layers.MaxPooling2D(pool_size=(2, 2)),
    tf.keras.layers.Conv2D(64, kernel_size=(3, 3),
activation='relu', name = 'conv_2',
      kernel_initializer='glorot_uniform', padding='same'),
    tf.keras.layers.MaxPooling2D(pool_size=(2, 2)),
    tf.keras.layers.Flatten(name = 'flat_1'),
    tf.keras.layers.Dense(256, activation='relu',
kernel_initializer='glorot_uniform', name = 'dense_64'),
    tf.keras.layers.Dense(10, activation='softmax',
name = 'custom_class')
  ])
  model.build([None, 32, 32, 3])

  model.compile(
    loss=tf.keras.losses.SparseCategoricalCrossentropy(
    from_logits=True),
    optimizer='adam',
    metrics=['accuracy'])
```

The remaining sections are the same as what you did in Chapter 7. You can define the directory name pattern to checkpoint the model during the training routine:

```python
MODEL_NAME = 'myCIFAR10-{}'.format(datetime.now().strftime(
"%Y%m%d-%H%M%S"))

checkpoint_dir = './' + MODEL_NAME
checkpoint_prefix = os.path.join(checkpoint_dir, "ckpt-{epoch}")
print(checkpoint_prefix)
```

The preceding command specifies the directory path to be something like ./ *myCIFAR10-20210302-014804/ckpt-{epoch}*.

Once you define the checkpoint directory, simply pass the definition to ModelCheckpoint. For simplicity, we'll only save the checkpoint if an epoch of training improves the model's accuracy on validation data over a previous epoch:

```
myCheckPoint = tf.keras.callbacks.ModelCheckpoint(
    filepath=checkpoint_prefix,
    monitor='val_accuracy',
    mode='max',
    save_weights_only=True,
    save_best_only = True)
```

Then wrap the definition around a list:

```
myCallbacks = [
    myCheckPoint
]
```

Now launch the training routine with the `fit` function, just like you have done in our other examples throughout the book:

```
hist = model.fit(
    train_dataset,
    epochs=12,
    steps_per_epoch=STEPS_PER_EPOCH,
    validation_data=validation_dataset,
    validation_steps=VALIDATION_STEPS,
    callbacks=myCallbacks).history
```

In the preceding command, the `hist` object contains information about the training results in a Python dictionary format. The property of interest in this dictionary is the item `val_accuracy`:

```
hist['val_accuracy']
```

This will display the validation accuracy from the first to the last epoch of training. From this list, we can determine the epoch with the highest accuracy in scoring the validation data. That's the model you want to use for scoring:

```
max_value = max(hist['val_accuracy'])
max_index = hist['val_accuracy'].index(max_value)
print('Best epoch: ', max_index + 1)
```

Since you set the checkpoint to save the best model rather than every epoch, a simpler alternative is to load the latest epoch:

```
tf.train.latest_checkpoint(checkpoint_dir)
```

This will give you the latest checkpoint under `checkpoint_dir`. Load the model with the weights from that checkpoint as:

```
model.load_weights(tf.train.latest_checkpoint(checkpoint_dir))
```

Use the model loaded with the best weights to score the test data:

```
eval_loss, eval_acc = model.evaluate(test_dataset)
print('Eval loss: {}, Eval Accuracy: {}'.format(
eval_loss, eval_acc))
```

Typical results look something like this:

```
1/1 [==============================] - 0s 726us/step
 loss: 1.7533
 accuracy: 0.7069
Eval loss: 1.753335952758789,
 Eval Accuracy: 0.706947386264801
```

Summary

What you have seen is the easiest way to get started with distributed TensorFlow model training. You learned how to create a cluster of GPUs from a commercial vendor's platform and how to refactor your single-node training code to a distributed training routine. In addition, you learned about the essentials of distributed machine learning and its different system architectures.

In the next section, you will learn another way to implement distributed training. You'll be using an open source library known as Horovod, created by Uber, which at its core also leverages the ring-allreduce algorithm. While this library requires more refactoring, it may serve as another option for you, in case you would like to compare the training time differences.

The Horovod API

In the preceding section, you learned how allreduce works. You also saw how the `tf.distribute` API automates facets of distributed training for you behind the scenes, so all you need to do is create a distributed training object and wrap the training code around the object's scope. In this section, you will learn about Horovod, an older distributed training API that requires you to handle these facets of distributed training in code. Since `tf.distribute` is popular and easy to use, the Horovod API isn't usually the first choice among programmers. The purpose of presenting it here is to give you another option for distributed training.

As in the previous section, we'll use Databricks as the platform for learning the basics of Horovod for distributed model training. If you followed my instructions, you will have a cluster consisting of two GPUs.

To understand how the Horovod API works, there are two critical parameters you need to know: each GPU's identity and the number of processes for parallel training. Each parameter is assigned to a Horovod environment variable, which will be used in your code. In this particular case, you should have two GPUs. Each will train on a shard of data, so there will be two training processes. You can retrieve Horovod environment variables using the following functions:

Rank
> Rank denotes a GPU's identity. If there are two GPUs, one GPU will be designated a rank value of 0 and the other a rank value of 1. And if there are more, the designated rank values will be 2, 3, and so forth.

Size
> Size denotes the total number of GPUs. If there are two, then Horovod's scope size is 2 and the training data will be split into two shards. Similarly, if there are four GPUs, this value will be 4 and the data will be split into four shards.

You will see these two functions being used very often. You can refer to the Horovod documentation (*https://oreil.ly/KC893*) for more details.

Code Pattern for Implementing the Horovod API

Before I show the full source code for running Horovod in Databricks, let's take a look at how to run a Horovod training job. The general pattern for running distributed training using Horovod in Databricks is:

```
from sparkdl import HorovodRunner
hr = HorovodRunner(np=2)
hr.run(train_hvd, checkpoint_path=CHECKPOINT_PATH,
learning_rate=LEARNING_RATE)
```

With Databricks, you need to run Horovod distributed training according to the preceding pattern. Basically, you create a HorovodRunner object called hr that allocates two GPUs. This object then executes a run function, which distributes a train_hvd function to each GPU. The train_hvd function is responsible for executing data ingestion and training routines at each GPU. Also, checkpoint_path is used to save the model at every training epoch, and learning_rate is used during the back propagation step of the training process.

As training proceeds through each epoch, the model weights and biases are aggregated, updated, and stored at the GPU ranked 0. learning_rate is another parameter designated by the Databricks driver and propagated to each GPU. Before you use the preceding pattern, though, you need to organize and implement several functions, which we'll go through next.

Encapsulating the Model Architecture

The job of Databricks' main driver is to distribute training data and model architecture blueprints to each GPU. Therefore, you need to wrap the model architecture in a function. As hr.run is executed, the train_hvd function is executed at each GPU. In train_hvd, a model architecture wrapper function such as this one is invoked:

```python
def get_model(num_classes):
  from tensorflow.keras import models
  from tensorflow.keras import layers

  model = tf.keras.Sequential([
  tf.keras.layers.Conv2D(32,
    kernel_size=(3, 3),
    activation='relu',
    name = 'conv_1',
    kernel_initializer='glorot_uniform',
    padding='same',
    input_shape = (32,32,3)),
  tf.keras.layers.MaxPooling2D(pool_size=(2, 2)),
  tf.keras.layers.Conv2D(64,
    kernel_size=(3, 3),
    activation='relu',
    name = 'conv_2',
    kernel_initializer='glorot_uniform',
    padding='same'),
  tf.keras.layers.MaxPooling2D(pool_size=(2, 2)),
  tf.keras.layers.Flatten(name = 'flat_1'),
  tf.keras.layers.Dense(256,
    activation='relu',
    kernel_initializer='glorot_uniform',
    name = 'dense_64'),
  tf.keras.layers.Dense(num_classes,
    activation='softmax',
    name = 'custom_class')
  ])
  model.build([None, 32, 32, 3])
  return model
```

As you can see, this is the same model architecture you used in the previous section, except it is wrapped as a function. The function will return the model object to the execution process in each GPU.

Encapsulating the Data Separation and Sharding Processes

To ensure each GPU receives a shard of training data, you also need to wrap the data processing steps as a function that can be passed into each GPU, just as the model architecture is passed.

As an example, let's use the same dataset, CIFAR-10, to illustrate how to ensure that each GPU gets a different shard of training data. Take a look at the following function:

```python
def get_dataset(num_classes, rank=0, size=1):
  from tensorflow.keras import backend as K
  from tensorflow.keras import datasets, layers, models
  from tensorflow.keras.models import Sequential
  import tensorflow as tf
  from tensorflow import keras
  import horovod.tensorflow.keras as hvd
  import numpy as np

  (train_images, train_labels), (test_images, test_labels) =
      datasets.cifar10.load_data()

  #50000 train samples, 10000 test samples.
  train_images = train_images[rank::size]
  train_labels = train_labels[rank::size]

  test_images = test_images[rank::size]
  test_labels = test_labels[rank::size]

  # Normalize pixel values to be between 0 and 1
  train_images, test_images = train_images / 255.0,
    test_images / 255.0

  return train_images, train_labels, test_images, test_labels
```

Notice the input parameters rank and size in the function signature. rank is defaulted to a value of 0, and size is defaulted to 1, so there is compatibility with single-node training. In distributed training with more than one GPU, each GPU will pass hvd.rank and hvd.size as inputs into this function. Since each GPU's identity is represented by hvd.rank through the double-colon (::) notation, images and labels are sliced and sharded according to how many steps to skip from one record to the next. As a result, the arrays returned by this function—train_images, train_labels, test_images, and test_labels—are different for each GPU, depending on its hvd.rank. (For a detailed explanation about NumPy array skipping and slicing, see this Colab notebook (*https://oreil.ly/23bmZ*).)

Parameter Synchronization Among Workers

It is important to initialize and synchronize the initial states of weights and biases among all workers (devices) before starting the training. This is done with a callback:

```
hvd.callbacks.BroadcastGlobalVariablesCallback(0)
```

This effectively broadcasts variable states from the 0-ranked GPU to all other GPUs.

Error metrics for all workers need to be averaged between each training step. This is done with another callback:

```
hvd.callbacks.MetricAverageCallback()
```

This is also passed into a callback list during training.

It's best to use a low learning rate early on and then switch to a preferred learning rate after, say, the first five epochs, which you can do by specifying the number of warm-up epochs with the following code:

```
hvd.callbacks.LearningRateWarmupCallback(warmup_epochs=5)
```

Also, include a way to reduce the learning rate during training when the model metric stops improving:

```
tf.keras.callbacks.ReduceLROnPlateau(patience=10, factor = 0.2)
```

In this example, you'll start to reduce the learning rate by a factor of 0.2 if there is no improvement in the model metric after 10 epochs.

To make things simpler, I recommend putting all these callbacks together as a list:

```
callbacks = [
   hvd.callbacks.BroadcastGlobalVariablesCallback(0),
   hvd.callbacks.MetricAverageCallback(),
   hvd.callbacks.LearningRateWarmupCallback(warmup_epochs=5,
     verbose=1),
   tf.keras.callbacks.ReduceLROnPlateau(patience=10, verbose=1)
]
```

Model Checkpoint as a Callback

As mentioned, after all workers complete an epoch of training, the model parameters are saved as a checkpoint in the 0-ranked worker. This is done using the following code snippet:

```
if hvd.rank() == 0:
   callbacks.append(keras.callbacks.ModelCheckpoint(
   filepath=checkpoint_path,
   monitor='val_accuracy',
   mode='max',
   save_best_only = True
   ))
```

This is necessary to prevent conflicts between workers, so that there is only one version of truth when it comes to model performance and validation metrics. As shown in the preceding code, with `save_best_only` set to True, the model and trained parameters will be saved only if the validation metric in that epoch is an improvement over the previous epoch. Therefore, not all epochs will result in a model being saved, and you can be sure that the latest checkpoint is the best model.

Distributed Optimizer for Gradient Aggregation

The gradient computation is also distributed, as each worker does its own training routine and calculates the gradient individually. You need to aggregate and then average all the gradients from different workers, then apply the average to all workers for the next step of training. This is accomplished with the following code snippet:

```
optimizer = tf.keras.optimizers.Adadelta(
lr=learning_rate * hvd.size())
optimizer = hvd.DistributedOptimizer(optimizer)
```

Here, `hvd.DistributedOptimizer` wraps the single-node optimizer's signature in Horovod's scope.

Distributed Training Using the Horovod API

Now let's take a look at a full implementation of distributed training using the Horovod API in Databricks. This implementation uses the same dataset (CIFAR-10) and model architecture you saw in "Using the Class tf.distribute.Mirrored-Strategy" on page 158:

```
import tensorflow as tf
import horovod.tensorflow.keras as hvd
import os
import time

def get_dataset(num_classes, rank=0, size=1):
  from tensorflow.keras import backend as K
  from tensorflow.keras import datasets, layers, models
  from tensorflow.keras.models import Sequential
  import tensorflow as tf
  from tensorflow import keras
  import horovod.tensorflow.keras as hvd
  import numpy as np

  (train_images, train_labels), (test_images, test_labels) =
datasets.cifar10.load_data()
  #50000 train samples, 10000 test samples.
  train_images = train_images[rank::size]
  train_labels = train_labels[rank::size]

  test_images = test_images[rank::size]
  test_labels = test_labels[rank::size]

  # Normalize pixel values to be between 0 and 1
  train_images, test_images = train_images / 255.0,
    test_images / 255.0

  return train_images, train_labels, test_images, test_labels
```

The preceding code will be executed at each worker. Each worker receives its own `train_images`, `train_labels`, `test_images`, and `test_labels`.

The following code is a function that wraps the model architecture; it will be built into each worker:

```
def get_model(num_classes):
  from tensorflow.keras import models
  from tensorflow.keras import layers
```

```
model = tf.keras.Sequential([
tf.keras.layers.Conv2D(32,
  kernel_size=(3, 3),
  activation='relu',
  name = 'conv_1',
  kernel_initializer='glorot_uniform',
  padding='same',
  input_shape = (32,32,3)),
tf.keras.layers.MaxPooling2D(pool_size=(2, 2)),
tf.keras.layers.Conv2D(64, kernel_size=(3, 3),
  activation='relu',
  name = 'conv_2',
  kernel_initializer='glorot_uniform',
  padding='same'),
tf.keras.layers.MaxPooling2D(pool_size=(2, 2)),
tf.keras.layers.Flatten(name = 'flat_1'),
tf.keras.layers.Dense(256, activation='relu',
  kernel_initializer='glorot_uniform',
  name = 'dense_64'),
tf.keras.layers.Dense(num_classes,
  activation='softmax',
  name = 'custom_class')
])
model.build([None, 32, 32, 3])
return model
```

Next is the main training function, train_hvd, which invokes the two functions just shown. This function is rather lengthy, so I'll explain it in blocks.

Inside train_hvd, a Horovod object is created and initialized with the command hvd.init. This function takes checkpoint_path and learning_rate as inputs for the distributed training routine to store models at each epoch and set the rate for gradient descent during the back propagation process. In the beginning, all libraries are imported:

```
def train_hvd(checkpoint_path, learning_rate=1.0):

    # Import tensorflow modules to each worker
    from tensorflow.keras import backend as K
    from tensorflow.keras.models import Sequential
    import tensorflow as tf
    from tensorflow import keras
    import horovod.tensorflow.keras as hvd
    import numpy as np
```

Then you create and initialize a Horovod object and use it to access your workers' configurations, so the data can be sharded properly later:

```
# Initialize Horovod
hvd.init()

# Pin GPU to be used to process local rank (one GPU per process)
# These steps are skipped on a CPU cluster
gpus = tf.config.experimental.list_physical_devices('GPU')
for gpu in gpus:
  tf.config.experimental.set_memory_growth(gpu, True)
if gpus:
  tf.config.experimental.set_visible_devices(
  gpus[hvd.local_rank()], 'GPU')

print(' Horovod size (processes): ', hvd.size())
```

Now that you've created your hvd object, use it to provide worker identification (hvd.rank) and the number of parallel processes (hvd.size) to the get_dataset function, which will return the training and validation data in shards.

Once you have these shards, convert them to a dataset so you can stream the training data, just as you did in "Using a GPU Cluster with tf.distribute.MirroredStrategy" on page 164:

```
# Call the get_dataset function you created, this time with the
  Horovod rank and size
num_classes = 10
train_images, train_labels, test_images, test_labels = get_dataset(
num_classes, hvd.rank(), hvd.size())

validation_dataset = tf.data.Dataset.from_tensor_slices(
  (test_images, test_labels))
train_dataset = tf.data.Dataset.from_tensor_slices(
  (train_images, train_labels))
```

Shuffle and batch the training and validation datasets:

```
NUM_CLASSES = len(np.unique(train_labels))
BUFFER_SIZE = 10000
BATCH_SIZE_PER_REPLICA = 64
validation_dataset_size = len(test_labels)
BATCH_SIZE = BATCH_SIZE_PER_REPLICA * hvd.size()
train_dataset = train_dataset.repeat().shuffle(BUFFER_SIZE).
```

```
    batch(BATCH_SIZE
  validation_dataset = validation_dataset.
  repeat().shuffle(BUFFER_SIZE).
batch(BATCH_SIZE, drop_remainder = True)

  train_dataset_size = len(train_labels)

  print('Training data sample size: ', train_dataset_size)

  validation_dataset_size = len(test_labels)
  print('Validation data sample size: ', validation_dataset_size)
```

Now define batch size, training steps, and training epochs:

```
TRAIN_DATASET_SIZE = len(train_labels)
STEPS_PER_EPOCH = TRAIN_DATASET_SIZE // BATCH_SIZE_PER_REPLICA
VALIDATION_STEPS = validation_dataset_size //
  BATCH_SIZE_PER_REPLICA
EPOCHS = 20
```

Create a model using the get_model function, set the optimizer, designate a learning rate, and then compile the model with the proper loss function for this classification task. Notice that the optimizer is wrapped by DistributedOptimizer for distributed training:

```
model = get_model(10)

# Adjust learning rate based on number of GPUs
optimizer = tf.keras.optimizers.Adadelta(
            lr=learning_rate * hvd.size())

# Use the Horovod Distributed Optimizer
optimizer = hvd.DistributedOptimizer(optimizer)

model.compile(optimizer=optimizer,
            loss=tf.keras.losses.SparseCategoricalCrossentropy(
            from_logits=True),
            metrics=['accuracy'])
```

Here you will create a callback list to synchronize variables across workers, to aggregate and average gradients for synchronous updates, and to adjust the learning rate according to epochs or training performance:

```
# Create a callback to broadcast the initial variable states from
  rank 0 to all other processes.
# This is required to ensure consistent initialization of all
```

```
# workers when training is started with random weights or
# restored from a checkpoint.
callbacks = [
    hvd.callbacks.BroadcastGlobalVariablesCallback(0),
    hvd.callbacks.MetricAverageCallback(),
    hvd.callbacks.LearningRateWarmupCallback(warmup_epochs=5,
      verbose=1),
    tf.keras.callbacks.ReduceLROnPlateau(patience=10,
      verbose=1)
  ]
```

Finally, here is the callback for model checkpoints at each epoch. This callback is only executed in the 0-ranked worker (hvd.rank() == 0):

```
# Save checkpoints only on worker 0 to prevent conflicts between
# workers
if hvd.rank() == 0:
    callbacks.append(keras.callbacks.ModelCheckpoint(
        filepath=checkpoint_path,
    monitor='val_accuracy',
    mode='max',
    save_best_only = True
    ))
```

Now comes the final `fit` function that will launch the model training routine:

```
model.fit(train_dataset,
            batch_size=BATCH_SIZE,
            epochs=EPOCHS,
            steps_per_epoch=STEPS_PER_EPOCH,
            callbacks=callbacks,
            validation_data=validation_dataset,
            validation_steps=VALIDATION_STEPS,
          verbose=1)
print('DISTRIBUTED TRAINING DONE')
```

This concludes the `train_hvd` function.

In the next cell of your Databricks notebook, specify a checkpoint directory for each epoch of training:

```
# Create directory
checkpoint_dir = '/dbfs/ml/CIFAR10DistributedDemo/train/{}/'.
format(time.time())
os.makedirs(checkpoint_dir)
print(checkpoint_dir)
```

checkpoint_dir will look something like */dbfs/ml/ CIFAR10DistributedDemo/train/1615074200.2146788/*.

In the next cell, go ahead and launch the distributed training routine:

```
from sparkdl import HorovodRunner

checkpoint_path = checkpoint_dir + '/checkpoint-{epoch}.ckpt'
learning_rate = 0.1
hr = HorovodRunner(np=2)
hr.run(train_hvd, checkpoint_path=checkpoint_path,
       learning_rate=learning_rate)
```

In the runner definition, HorovodRunner(np=2), the number of processes is specified as two per setup (see "Setting Up Distributed Training" on page 160), which sets up two Standard_NC12 worker GPUs.

Once the training routine is complete, take a look at the checkpoint directories using the following command:

```
ls -lrt /dbfs/ml/CIFAR10DistributedDemo/train/1615074200.2146788/
```

You should see something like this:

```
drwxrwxrwx 2 root root 4096 Mar 6 23:18 checkpoint-9.ckpt/
drwxrwxrwx 2 root root 4096 Mar 6 23:18 checkpoint-8.ckpt/
drwxrwxrwx 2 root root 4096 Mar 6 23:18 checkpoint-7.ckpt/
drwxrwxrwx 2 root root 4096 Mar 6 23:18 checkpoint-6.ckpt/
drwxrwxrwx 2 root root 4096 Mar 6 23:18 checkpoint-5.ckpt/
drwxrwxrwx 2 root root 4096 Mar 6 23:18 checkpoint-4.ckpt/
drwxrwxrwx 2 root root 4096 Mar 6 23:18 checkpoint-3.ckpt/
drwxrwxrwx 2 root root 4096 Mar 6 23:18 checkpoint-2.ckpt/
drwxrwxrwx 2 root root 4096 Mar 6 23:18 checkpoint-20.ckpt/
drwxrwxrwx 2 root root 4096 Mar 6 23:18 checkpoint-1.ckpt/
drwxrwxrwx 2 root root 4096 Mar 6 23:18 checkpoint-19.ckpt/
drwxrwxrwx 2 root root 4096 Mar 6 23:18 checkpoint-17.ckpt/
drwxrwxrwx 2 root root 4096 Mar 6 23:18 checkpoint-16.ckpt/
drwxrwxrwx 2 root root 4096 Mar 6 23:18 checkpoint-15.ckpt/
drwxrwxrwx 2 root root 4096 Mar 6 23:18 checkpoint-14.ckpt/
drwxrwxrwx 2 root root 4096 Mar 6 23:18 checkpoint-13.ckpt/
drwxrwxrwx 2 root root 4096 Mar 6 23:18 checkpoint-11.ckpt/
drwxrwxrwx 2 root root 4096 Mar 6 23:18 checkpoint-10.ckpt/
```

Some checkpoints are skipped if there is no model improvement over previous epochs. The latest checkpoint represents the model with the best validation metrics.

Wrapping Up

In this chapter, you learned what it takes for distributed model training to work in an environment with multiple workers. With a data parallelism framework, there are two major patterns for distributed training: asynchronous parameter server and synchronous allreduce. Today synchronous allreduce is more popular because of the general availability of high-performance accelerators.

You learned how to use a Databricks GPU cluster to perform two types of synchronous allreduce APIs: TensorFlow's own `tf.distribute` API and Uber's Horovod API. The TensorFlow option provides the most elegant and convenient use and requires the least amount of code refactoring, whereas the Horovod API requires users to manually handle data sharding, distribution pipelines, gradient aggregation and averaging, and model checkpoints. Both options perform distributed training by ensuring that each worker performs its own training, and then, at the end of each training step, updating the gradients synchronously and consistently among all workers. This is the hallmark of distributed training.

Congratulations—by working your way through this chapter, you learned how to train a deep-learning model with a distributed data pipeline and distributed training routine, using a cluster of GPUs in the cloud. In the next chapter, you will learn how to serve a TensorFlow model for inferencing.

## CHAPTER 9

Serving TensorFlow Models

If you've been reading the chapters in this book sequentially, you now know a lot about how to handle the data engineering pipeline, build models, launch training routines, checkpoint models at each epoch, and even score test data. In all the examples thus far, these tasks have mostly been wrapped together for didactic purposes. In this chapter, however, you're going to learn more about how to serve TensorFlow models based on the format in which they are saved.

Another important distinction between this chapter and previous chapters is that here you will learn a coding pattern for handling data engineering for test data. Previously, you saw that test data and training data are transformed at the same runtime. As a machine learning engineer, though, you also have to think about the scenarios where your model is deployed.

Imagine that your model is loaded in a Python runtime and ready to go. You have a batch of samples or a sample. What do you need to do to the input data so that the model can accept it and return predictions? In other words: you have a model and raw test data; how do you implement the logic of transforming the raw data? In this chapter, you will learn about serving the model through a few examples.

Model Serialization

A TensorFlow model can be saved in two different native formats (without any optimization): HDF5 (h5) or protobuf (pb). Both formats are standard data serialization (saving) formats in Python and other programming languages to persist objects or data structures; they are not specific to TensorFlow or even ML models. Before TensorFlow 2.0, pb was the only native format available. With TensorFlow 2.0, where the Keras API is the de facto high-level API going forward, h5 has become an alternative to pb.

Today both formats may be used for deployment, especially in various public cloud providers. This is because each cloud provider now has its own API that wraps a model. As long as the model is saved in your workspace, you can reach it through web services such as RESTful (Representational State Transfer) APIs, which utilize HTTP methods to make a request over a network. Therefore, regardless of format, your model is ready for serving through a RESTful API call from a client program.

Let's start with the image classification model you built with CIFAR-10 data in the last chapter. If you haven't worked your way through that chapter, use the following code snippets to get an image classifier built and trained quickly. (By the way, the code here was developed in Google Colab with one GPU.)

Import all the necessary libraries first:

```
import tensorflow as tf
from tensorflow.keras import datasets, layers, models
import numpy as np
import matplotlib.pylab as plt
import os
from datetime import datetime
```

Load and normalize the images:

```
(train_images, train_labels), (test_images, test_labels) =
datasets.cifar10.load_data()

train_images, test_images = train_images / 255.0,
 test_images / 255.0
```

Define your image labels:

```
CLASS_NAMES = ['airplane', 'automobile', 'bird', 'cat',
               'deer','dog', 'frog', 'horse', 'ship', 'truck']
```

Transform the raw images into a dataset. Since a partition for test images is available, the following code separates out the first 500 images in that partition for the validation dataset used at the end of each training epoch, and keeps the remainder as a test dataset. All of the training images are converted to training datasets:

```
validation_dataset = tf.data.Dataset.from_tensor_slices(
(test_images[:500], test_labels[:500]))

test_dataset = tf.data.Dataset.from_tensor_slices(
(test_images[500:], test_labels[500:]))

train_dataset = tf.data.Dataset.from_tensor_slices(
(train_images, train_labels))
```

To make sure you know the sample size of each dataset, iterate through them and display the sample size:

```
train_dataset_size = len(list(train_dataset.as_numpy_iterator()))
print('Training data sample size: ', train_dataset_size)

validation_dataset_size = len(list(validation_dataset.
as_numpy_iterator()))
print('Validation data sample size: ', validation_dataset_size)

test_dataset_size = len(list(test_dataset.as_numpy_iterator()))
print('Test data sample size: ', test_dataset_size)
```

You should see the following output:

```
Training data sample size:  50000
Validation data sample size:  500
Test data sample size:  9500
```

Define a distribution strategy for distributed training:

```
strategy = tf.distribute.MirroredStrategy()
print('Number of devices: {}'.format(
strategy.num_replicas_in_sync))
```

You should see one GPU available:

```
Number of devices: 1
```

Now set the batch sizes for training:

```
BUFFER_SIZE = 10000
BATCH_SIZE_PER_REPLICA = 64
BATCH_SIZE = BATCH_SIZE_PER_REPLICA * strategy.num_replicas_in_sync
```

Apply the batches to each dataset and calculate the number of batches for each training epoch before evaluating the model's accuracy with the validation data:

```
train_dataset = train_dataset.repeat().shuffle(BUFFER_SIZE).batch(
                BATCH_SIZE)

validation_dataset = validation_dataset.shuffle(BUFFER_SIZE).batch(
                validation_dataset_size)

test_dataset = test_dataset.batch(test_dataset_size)

STEPS_PER_EPOCH = train_dataset_size // BATCH_SIZE_PER_REPLICA
VALIDATION_STEPS = 1
```

Create a function called `build_model` to define the model architecture and compile it with the loss function for classification, optimizer, and metrics, all within the distributed training strategy scope:

```
def build_model():
  with strategy.scope():
    model = tf.keras.Sequential([
      tf.keras.layers.Conv2D(32, kernel_size=(3, 3),
        activation='relu',
        name = 'conv_1',
        kernel_initializer='glorot_uniform',
        padding='same',
        input_shape = (32,32,3)),
      tf.keras.layers.MaxPooling2D(pool_size=(2, 2)),
      tf.keras.layers.Conv2D(64, kernel_size=(3, 3),
        activation='relu',
        name = 'conv_2',
        kernel_initializer='glorot_uniform',
        padding='same'),
      tf.keras.layers.MaxPooling2D(pool_size=(2, 2)),
      tf.keras.layers.Flatten(name = 'flat_1'),
      tf.keras.layers.Dense(256, activation='relu',
kernel_initializer='glorot_uniform', name = 'dense_64'),
      tf.keras.layers.Dense(10, activation='softmax',
        name = 'custom_class')
    ])
```

```
model.build([None, 32, 32, 3])

model.compile(
  loss=tf.keras.losses.SparseCategoricalCrossentropy(
      from_logits=True),
  optimizer=tf.keras.optimizers.Adam(),
  metrics=['accuracy'])
return model
```

Now create a model instance by invoking `build_model`:

```
model = build_model()
```

Define an alias for the file path and to save model checkpoints:

```
MODEL_NAME = 'myCIFAR10-{}'.format(datetime.now().strftime(
"%Y%m%d-%H%M%S"))
```

Use a `print` statement to display the model name format:

```
print(MODEL_NAME)
myCIFAR10-20210319-214456
```

Now set up your checkpoint directory:

```
checkpoint_dir = './' + MODEL_NAME
checkpoint_prefix = os.path.join(checkpoint_dir,
    "ckpt-{epoch}")
print(checkpoint_prefix)
```

The checkpoint directory will be set as a directory with the following pattern:

```
./myCIFAR10-20210319-214456/ckpt-{epoch}
```

In your current directory level, you will see a
myCIFAR10-20210319-214456 directory containing weight files
with the prefix *ckpt-{epoch}*, where *{epoch}* is the epoch
number.

Next, define a checkpoint object. Let's save the model weights
at the end of an epoch only if the model performance on validation data is improved over the previous epoch. Save the
weights in the same directory (*myCIFAR10-20210319-214456*)
so that the latest saved checkpoint weights are from the best
epoch. This saves time, since you don't need to determine
which epoch presented the best model. Make sure that both
`save_weights_only` and `save_best_only` are set to True:

```
myCheckPoint = tf.keras.callbacks.ModelCheckpoint(
    filepath=checkpoint_prefix,
    monitor='val_accuracy',
    mode='max',
    save_weights_only = True,
    save_best_only = True
    )
```

Now pass the preceding checkpoint definition into a list, as required by the fit function:

```
myCallbacks = [
myCheckPoint
]
```

And launch the training process:

```
model.fit(
    train_dataset,
    epochs=30,
    steps_per_epoch=STEPS_PER_EPOCH,
    validation_data=validation_dataset,
    validation_steps=VALIDATION_STEPS,
    callbacks=myCallbacks)
```

Throughout the training routine, you can inspect the output. You'll likely see something similar to this:

```
Epoch 7/30
781/781 [==============================] - 4s 6ms/step
loss: 0.0202 - accuracy: 0.9972 - val_loss: 10.5573
val_accuracy: 0.6900
Epoch 8/30
781/781 [==============================] - 4s 6ms/step
loss: 0.0217 - accuracy: 0.9967 - val_loss: 10.4517
val_accuracy: 0.7000
Epoch 9/30
781/781 [==============================] - 5s 6ms/step
loss: 0.0203 - accuracy: 0.9971 - val_loss: 10.4553
val_accuracy: 0.7080
Epoch 10/30
781/781 [==============================] - 5s 6ms/step
loss: 0.0232 - accuracy: 0.9966 - val_loss: 11.3774
val_accuracy: 0.6600
…
Epoch 30/30
781/781 [==============================] - 5s 6ms/step
loss: 0.0221 - accuracy: 0.9971 - val_loss: 11.9106
val_accuracy: 0.6680
<tensorflow.python.keras.callbacks.History at 0x7fd447ce9a50>
```

In this example, the highest validation accuracy occurs in epoch 9, where `val_accuracy` is 0.7080.

Check the checkpoint directory:

```
!ls -lrt {checkpoint_dir}
```

You will see its contents:

```
total 87896
-rw-r--r-- 1 root root 12852661 Mar 19 21:45
ckpt-1.data-00000-of-00001
-rw-r--r-- 1 root root     2086 Mar 19 21:45 ckpt-1.index
-rw-r--r-- 1 root root 12852661 Mar 19 21:45
ckpt-2.data-00000-of-00001
-rw-r--r-- 1 root root     2086 Mar 19 21:45 ckpt-2.index
-rw-r--r-- 1 root root 12852661 Mar 19 21:45
ckpt-3.data-00000-of-00001
-rw-r--r-- 1 root root     2086 Mar 19 21:45 ckpt-3.index
-rw-r--r-- 1 root root 12852661 Mar 19 21:45
ckpt-4.data-00000-of-00001
-rw-r--r-- 1 root root     2086 Mar 19 21:45 ckpt-4.index
-rw-r--r-- 1 root root 12852661 Mar 19 21:46
ckpt-7.data-00000-of-00001
-rw-r--r-- 1 root root     2086 Mar 19 21:46 ckpt-7.index
-rw-r--r-- 1 root root 12852661 Mar 19 21:46
ckpt-8.data-00000-of-00001
-rw-r--r-- 1 root root     2086 Mar 19 21:46 ckpt-8.index
-rw-r--r-- 1 root root 12852661 Mar 19 21:46
ckpt-9.data-00000-of-00001
-rw-r--r-- 1 root root     2086 Mar 19 21:46 ckpt-9.index
-rw-r--r-- 1 root root       69 Mar 19 21:46 checkpoint
```

Since the `myCheckpoint` object has `save_best_only` and `save_weights_only` set to True, the last weight is `ckpt-9.data-00000-of-00001`.

To programmatically locate the last weight file among all the saved weight files from a single directory, you can use the `latest_checkpoint` function in the `tf.train` API. Run the following command:

```
tf.train.latest_checkpoint(checkpoint_dir)
```

This is the expected output:

```
./myCIFAR10-20210319-214456/ckpt-9
```

This identifies the prefix to the last weight file. You can then load the best weights to the model:

```
model.load_weights(tf.train.latest_checkpoint(checkpoint_dir))
```

Now you have the best model, which can be saved in h5 or pb format. We'll look at h5 first.

Saving a Model to h5 Format

The high-level `tf.keras` API uses the `save` function as a means to save the model in h5 format:

```
KERAS_MODEL_PATH = "/models/HDF5/tfkeras_cifar10.h5"
model.save(KERAS_MODEL_PATH)
```

Take a look at the directory:

```
!ls -lrt {KERAS_MODEL_PATH}
```

You will see the model as an h5 file:

```
-rw-r--r-- 1 root root 12891752 Mar 20 21:19
/models/HDF5/tfkeras_cifar10.h5
```

In the future, if you want to reload this model for scoring, simply use the `load_model` function. (Make sure you also import all the libraries, as indicated in the beginning of the section.)

```
new_h5_model = tf.keras.models.load_model(
'/models/HDF5/tfkeras_cifar10.h5')
```

For quick scoring, use `test_dataset`, which you prepared at the same time as `training_dataset`:

```
new_h5_model.predict(test_dataset)
```

It will produce results like these:

```
array([[1.77631108e-07, 8.12380506e-07, 9.94834751e-02, ...,
        4.93609463e-04, 1.97697682e-05, 2.55090754e-06],

       [6.76187535e-12, 6.38716233e-11, 1.67756411e-07, ...,
        9.99815047e-01, 1.25759464e-14, 1.24499985e-11]],
        dtype=float32)
```

There are 9,500 elements; each is an array of probabilities. The index of maximum probability maps to the label in CLASS_NAMES.

Saving a Model to pb Format

To save the same model to pb format, you'll use the tf.saved_model.save function:

```
SAVED_MODEL_PATH = "/models/pb/tfsaved_cifar10"
tf.saved_model.save(model, SAVED_MODEL_PATH)
```

Look at the contents in SAVED_MODEL_PATH:

```
!ls -lrt {SAVED_MODEL_PATH}
```

The contents should look like this:

```
drwxr-xr-x 2 root root   4096 Mar 20 21:50 variables
drwxr-xr-x 2 root root   4096 Mar 20 21:50 assets
-rw-r--r-- 1 root root 138184 Mar 20 21:50 saved_model.pb
```

The weight file is in the variables directory. You can inspect it with this command:

```
!ls -lrt {SAVED_MODEL_PATH}/variables
```

Here's what you'll see (more or less):

```
-rw-r--r-- 1 root root 12856259 Mar 20 21:50
variables.data-00000-of-00001
-rw-r--r-- 1 root root     2303 Mar 20 21:50 variables.index
```

Now that you've seen the folder structure when a model is saved as a protobuf, let's see how to load a model protobuf. In this case, you'll need to load it from the directory name, which is the directory that contains *saved_model.pb*:

```
load_strategy = tf.distribute.MirroredStrategy()
with load_strategy.scope():
  load_options = tf.saved_model.LoadOptions(
             experimental_io_device='/job:localhost')
  loaded_pb = tf.keras.models.load_model(
             SAVED_MODEL_PATH,
             options=load_options)
```

If you take a closer look at the preceding commands, you will notice that just as in model training, model loading is done within a distribute strategy scope. If you are running a cloud TPU or GPU (as in the case of Google Colab (*https://oreil.ly/ZBYwr*)), set experimental_io_device to localhost, which is the node where you saved the model. Then use tf.keras.mod els.load_model to load the directory that holds *saved_model.pb*: in this case, it is SAVED_MODEL_PATH.

Now use the model loaded_pb to score test_dataset:

```
loaded_pb.predict(test_dataset)
```

You will see the same output as in the h5 model's predictions:

```
array([[1.77631108e-07, 8.12380506e-07, 9.94834751e-02, ...,
        4.93609463e-04, 1.97697682e-05, 2.55090754e-06],

       [6.76187535e-12, 6.38716233e-11, 1.67756411e-07, ...,
        9.99815047e-01, 1.25759464e-14, 1.24499985e-11]],
dtype=float32)
```

Likewise, each inner bracket is a list of probabilities for a test image. The index of maximum probability can be mapped to the correct entry in CLASS_NAMES.

As you've seen, the model in either h5 or pb format can be used for scoring test data in dataset format. The model can also score test data in a NumPy array format. Recall that test_images[500:] is the original NumPy test data format; the subset starts at 500 images and goes on (for a total of 9,500 test images). You can pass this NumPy array directly into the model for scoring:

```
loaded_pb.predict(test_images[500:])
```

You will see the same output:

```
array([[1.77631108e-07, 8.12380506e-07, 9.94834751e-02, ...,
        4.93609463e-04, 1.97697682e-05, 2.55090754e-06],

       [6.76187535e-12, 6.38716233e-11, 1.67756411e-07, ...,
        9.99815047e-01, 1.25759464e-14, 1.24499985e-11]],
dtype=float32)
```

Selecting the Model Format

You have now seen how to score test data with both the h5 and pb model formats. However, choosing which format to use depends on many things. Conceptually, the h5 format is very easy to understand; it consists of a model skeleton and weights, saved as a single file. This is very similar to how a pickle object or file works: as long as you import the library, you can open the single file that contains everything you need to reinstate the object (in this case, your model). This approach is suitable for simple deployments, where a driver program running the Python runtime can simply use tf.keras.models.load_model to load the model and run it over the test data.

However, if the model has to be run with more complicated settings, then protobuf format is a better choice. This is because the pb format is programming-language agnostic: it can be read by many other programming languages besides Python, such as Java, JavaScript, C, C++, and so forth. In fact, when you take the model to production, you will use TensorFlow Serving to host the pb model to score test data over the internet. In the next section, you will learn how TensorFlow Serving works.

TensorFlow Serving

TensorFlow Serving (TFS) is a framework specifically designed for running ML models in a production environment. Since scoring test data over the internet (or using Internet Protocol in a virtual private network) is arguably the most common model-serving scenario, there needs to be an HTTP or HTTPS endpoint that serves as the "front door" to the model. The client program, which will pass test data to the model, needs to communicate with the model's endpoint via HTTP. This communication follows the style of the RESTful API, which specifies a set of rules and formats for data sent over HTTP.

TFS takes care of all the complexity here for you. Next you will see how to run TFS to host this model in your local environment.

Running TensorFlow Serving with a Docker Image

The easiest way to learn TFS is with a Docker image. If you need some background on Docker and general container technology, take a look at *Using Docker* (*https://oreil.ly/FyJOH*) by Adrian Mouat (O'Reilly). Chapter 1 provides a concise explanation of Docker containers, while Chapter 2 shows you how to install a Docker engine and get it up and running in your local node.

Briefly, a Docker image is a lightweight, standalone, executable package of software that includes everything needed to run an application: code, runtime, system tools, system libraries, and settings. To run a Docker image, you need a Docker engine. When you run a Docker image on a Docker engine, the image becomes a *container* (*https://oreil.ly/V8KMi*).

For instructions on installing a Docker engine, take a look at the Docker documentation (*https://oreil.ly/77l6U*). There are versions available for macOS, Windows 10, and Linux. Choose the one that works for your environment and follow the installation instructions. For the examples and workflow in this chapter, my local system is running macOS Big Sur version 11.2, and my Docker engine version is 3.0.3.

Now make sure your Docker engine is up and running: launch it by double-clicking its icon in your environment. When it's running, you will see the Docker whale icon in the top bar on a Mac, shown in Figure 9-1 or in the notification area (lower right) on a PC.

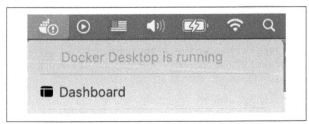

Figure 9-1. Docker engine running status

Once you have your Docker engine installed and running, download TFS's Docker image as a base, add the CIFAR-10 model to the base image, and then build a new image. This new image will be served through an HTTP endpoint and a specific TCP/IP port. A client program will send test data to this HTTP address and port.

Make sure you save your model in pb format. This time, name it *001*. This directory doesn't have to be named *001*, but it does have to be numeric per TFS's required hierarchy and naming convention.

Continue with the notebook you made in the previous section, and save the model in the local directory by using the following command:

```
SAVED_MODEL_PATH = "./models/CIFAR10/001"
tf.saved_model.save(model, SAVED_MODEL_PATH)
```

This will produce the following directory structure:

```
models
    CIFAR10
        001
            assets
            saved_model.pb
            variables
```

For now, use the command terminal and navigate to the *models* directory.

Now that the Docker engine is running, you are ready to start pulling a TFS image. Type the following command while in the *models* directory:

```
docker pull tensorflow/serving
```

This command downloads a TFS image to your local Docker environment. Now run the image:

```
docker run -d --name serv_base_img tensorflow/serving
```

The preceding command launches a TFS image as a container named `serv_base_img`. Run the following command to add the model you built to the base image:

```
docker cp $PWD/CIFAR10 serv_base_img:/models/cifar10
```

It is natural to think of the *saved_model.pb* file as a reference for where everything else is. Remember that *CIFAR10* is the local directory, two levels up from the pb file. In between them is the directory *001*. Now *CIFAR10* is copied into the base image as */models/cifar10*. Notice that in Docker, directory names are all lowercase.

Next, commit the change you made to the base image. Run the following command:

```
docker commit --change "ENV MODEL_NAME cifar10model"
serv_base_img cifar10model\
```

Now you can stop the base image; you don't need it anymore:

```
docker kill serv_base_img
```

Let's review what you've done so far. You have created a new Docker image by adding your CIFAR-10 model to TFS, which is a base image. That model is now deployed and running in the TFS container. Once you run the TFS container, the model is live and ready to serve any client.

To serve the TFS container that hosts your CIFAR-10 model, run the following command:

```
docker run -p 8501:8501 \
  --mount type=bind,\
source=$PWD/CIFAR10,\
target=/models/cifar10 \
  -e MODEL_NAME=cifar10 -t tensorflow/serving &
```

This is a relatively lengthy command. Let's dissect it a bit.

First, you map local port 8501 to the Docker engine's port 8501. There is nothing magical about your local port number. If 8501 is in use in your local environment, you can use a different port number—say, 8515. If so, then the command would be -p 8515:8501. Since the TFS container always runs on port 8501, the second target in the preceding command is always 8501.

The source indicates that below the current directory (*$PWD*) there is a *CIFAR10* directory, which is where the model is located. This model is named CIFAR10, and the `tensorflow/serv`ing container is ready to take input.

You will see the output shown in Figure 9-2. It indicates that you are running CIFAR10 model version 1, which is taken from the directory named *001*.

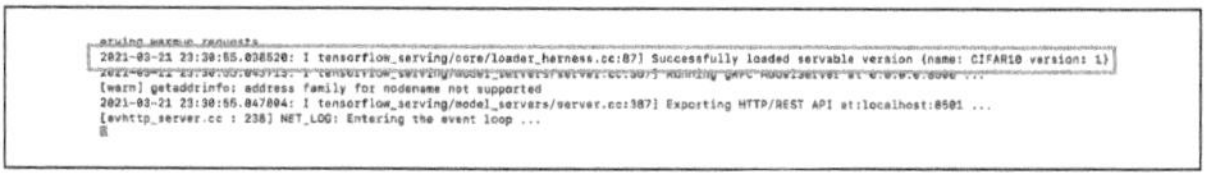

Figure 9-2. Command terminal running a custom Docker container

Scoring Test Data with TensorFlow Serving

Now that your TFS container is running with your model, you are ready to pass test data to the model. This is done via an HTTP request. You may use another Jupyter Notebook as a client that sends the NumPy array to TFS. The HTTP address for TFS is *http://localhost:8501/v1/models/cifar10:predict*.

Here is the client code in a different Jupyter Notebook:

1. Import all the necessary libraries:

```
import tensorflow as tf
from tensorflow.keras import datasets
import requests
import json
import numpy as np
```

2. Load test images and normalize the pixel value range to be between 0 and 1, and then select the images. For simplicity, let's only select 10 images; we'll use those that are between 500 and 510:

```
(train_images, train_labels), (test_images, test_labels) =
datasets.cifar10.load_data()

# Normalize pixel values to be between 0 and 1
train_images, test_images = train_images / 255.0,
 test_images / 255.0
test_images = test_images[500:510]
```

3. Convert the NumPy array `test_images` into JSON, a commonly used format for data exchange between a client and a server over HTTP:

```
DATA = json.dumps({
    "instances": test_images.tolist()
})
```

You also need to define headers:

```
HEADERS = {"content-type": "application/json"}:
```

Now that you have wrapped your NumPy data with the appropriate format and headers, you are ready to send the whole package to TFS.

4. Construct the entire package as an HTTP request:

```
response = requests.post(
'http://localhost:8501/v1/models/cifar10:predict',
data=DATA, headers=HEADERS)
```

Notice that you use the `post` method to solicit a response from TFS. TFS will score `DATA` and return the results as `response`. This communication framework, which uses a JSON request format and rules to establish and handle communication between a client and a server, is also known as a RESTful API.

5. Take a look at what TFS has predicted:

```
response.json()
```

The preceding command will decode the prediction to an array of probability:

```
{'predictions': [[9.83229938e-07,
    1.24386987e-10,
    0.0419323482,
    0.00232415553,
    0.91928196,
    3.26286099e-05,
    0.0276549552,
    0.00877290778,
```

```
    8.02750222e-08,
    5.4040652e-09],
....
[2.60355654e-10,
    5.17050935e-09,
    0.000181202529,
    1.92517109e-06,
    0.999798834,
    1.04122219e-05,
    3.32912987e-06,
    4.38272036e-06,
    4.2479078e-09,
    9.54967494e-11]]}
```

In the preceding code, you see the first and tenth test images. Each inner array consists of 10 probability values, each of which maps to a CLASS_NAME.

6. To map the maximum probability in each prediction back to a label, you need to retrieve the values, shown earlier, from the Python dictionary response. You can retrieve the values using the key name predictions via:

```
predictions_prob_list = response.json().get('predictions')
```

The labels for CIFAR-10 data are:

```
CLASS_NAMES = ['airplane', 'automobile', 'bird', 'cat',
               'deer','dog', 'frog', 'horse', 'ship', 'truck']
```

CLASS_NAMES is a list that holds the CIFAR-10 labels.

7. Now convert predictions_prob_list to a NumPy array, then use argmax to find the index for the maximum probability value:

```
predictions_array = np.asarray(predictions_prob_list)
predictions_idx = np.argmax(predictions_array, axis = 0)
```

8. Map each index (there are 10) to a CIFAR-10 label:

```
for i in predictions_idx:
    print(CLASS_NAMES[i])
```

Your output should look something like this:

```
ship
ship
airplane
bird
truck
ship
automobile
frog
ship
horse
```

This is how you decode the probability array back to labels.

You have just run a TFS Docker container with your own image classification model behind an HTTP endpoint. That container accepts input data as a JSON payload in a `post` request. TFS unpacks the request body; extracts the JSON payload, which contains the NumPy array; scores each array; and returns the results back to the client.

Wrapping Up

This chapter showed you the basics of model persistence (saving) and model serving (scoring). The TensorFlow model is flexible in that it takes advantage of the simplicity offered by the `tf.keras` API to save the model as a single HDF5 file. This format is easy to handle and share with others.

For a serving framework that caters to a production environment, typically you need to have a model hosted in a runtime, and that runtime needs to be accessible via a web-based communication protocol such as HTTP. As it turns out, TFS provides a framework that handles the HTTP request. All you need to do is copy your protobuf model folder to a TFS base image and commit the change to the base image. Now you have created a Docker image of your model, and you have the model running behind TFS.

You learned how to use another runtime to create a correctly shaped numeric array, wrap it around a JSON-format data payload, and send it using the `post` command to the HTTP endpoint hosted by TFS for scoring.

This pretty much completes the knowledge loop of building, training, and serving the model. In the next chapter, you will learn more practices for model tuning and fairness.

## CHAPTER 10

Improving the Modeling Experience: Fairness Evaluation and Hyperparameter Tuning

Getting ML models to work well is an iterative process. It requires many rounds of tuning the model parameters, architectures, and training durations. While you have to work with the data that's available, of course, ideally you want the training data to be balanced. In other words, it should contain an equal number of classes or uniform distribution across ranges.

Why is this balance important? Because if any features in the data are skewed, then the trained model will reproduce that skew. This is known as *model bias*.

Imagine that you're training a model to recognize dogs. If there are 99 negative cases and 1 positive case in your training images—that is, only one actual dog image—then the model will simply predict a negative result every time, with a 99% chance of being correct. The model learns to minimize the errors it makes during training, and the easiest way to do so is to produce a negative prediction—in short, to guess "not a dog" every time. This is known as the data imbalance problem, and it is prevalent in the real world; it's also a complicated subject to which I cannot do justice here. It requires many different approaches, including adding synthetic data through a technique known as *data augmentation.*

In this chapter, I'll introduce you to Fairness Indicators, a new tool (as of this writing) to evaluate model bias. It is part of the TensorFlow Model Analysis library and is available for Jupyter Notebooks.

You will also learn how to perform *hyperparameter tuning.* *Hyperparameters* are variables in the model architecture and model training process. Sometimes you want to experiment with different values or implementation choices, but you don't know which are best for your model. To find out, you'll need to evaluate model performance over many combinations of hyperparameters. I'll show you a new way of doing hyperparameter tuning using the Keras Tuner library. This library works seamlessly with TensorFlow 2's Keras API. It's very flexible and easy to set up as part of the training process. We'll start by setting up Fairness Indicators in the next section.

TIP

Model bias, and its real-life consequences, are well known. One of the most notable examples of model bias is the COMPAS (Correctional Offender Management Profiling for Alternative Sanctions) framework, which was used in US court systems to predict recidivism. Because of the training data, the model predicted twice as many false positives (*https://oreil.ly/1FdFy*) for Black defendants as for white defendants. If you are interested in learning more about fairness, take a look at *Practical Fairness* by Aileen Nielsen (*https://oreil.ly/V3yjB*) (O'Reilly, 2020) and *AI Fairness* by Trisha Mahoney, Kush R. Varshney, and Michael Hind (*https://oreil.ly/tpoTl*) (O'Reilly, 2020).

Model Fairness

You will need to install the TensorFlow Model Analysis library, which was not a part of the regular TensorFlow distribution as of TensorFlow 2.4.1. You can download and install it via the `pip install` command:

```
pip install tensorflow-model-analysis
```

You will also need to install the `protobuf` library to parse your choice of model metrics:

```
pip install protobuf
```

This library enables you to display and review model statistics on test data so that you can detect any bias in the model's prediction.

To illustrate this, we will again use the *Titanic* dataset. In Chapter 3, you used this dataset to build a model to predict passenger survival. This small dataset contains several features about each passenger and is a good starting point here.

We see survival as a discrete outcome: someone either survived or did not. However, for a model, what we really mean is the probability of survival based on the given features of a passenger. Recall that the model you built is a logistic regression model, and the output is a probability of an outcome that is binary (survived or not). One Google course (*https://oreil.ly/8ciQJ*) puts it this way:

> In order to map a logistic regression value to a binary category, you must define a *classification threshold* (also called the *decision threshold*). A value above that threshold indicates [a positive]; a value below indicates [a negative]. It is tempting to assume that the classification threshold should always be 0.5, but thresholds are problem-dependent, and are therefore values that you must tune.

It is up to the user to decide the threshold. An intuitive way to understand this is that a survival probability of 0.51 doesn't guarantee survival; it still implies a 49% chance of *not* surviving. Likewise, a survival probability of 0.49 is not zero. A good threshold is one that minimizes misclassification in both directions. Therefore, the threshold is a user-determined parameter. Typically, in your model training and testing process, you will try a few different threshold values and see which one gives you the most correct classifications in the test data. For this dataset, you might start with a list of different threshold values, such as 0.1, 0.5, and 0.9. For each threshold, a positive result—that is, a prediction probability that falls above the threshold—indicates a prediction that this individual survived.

Recall that the *Titanic* dataset looks like what is shown in Figure 10-1.

	survived	sex	age	n_siblings_spouses	parch	fare	class	deck	embark_town	alone
0	0	male	35.0	0	0	8.0500	Third	unknown	Southampton	y
1	0	male	54.0	0	0	51.8625	First	E	Southampton	y
2	1	female	58.0	0	0	26.5500	First	C	Southampton	y
3	1	female	55.0	0	0	16.0000	Second	unknown	Southampton	y
4	1	male	34.0	0	0	13.0000	Second	D	Southampton	y
...	...	...	...	...	...	...	...	...	...	...

Figure 10-1. Titanic dataset for training

Each row in Figure 10-1 represents a passenger and several corresponding features. The model's goal is to predict the value in the "survived" column based on these features. In training data, this column is binary, with 1 indicating the passenger survived and 0 indicating the passenger did not survive. Data for testing is also provided by the *Titanic* dataset as a separate partition.

Model Training and Scoring

Let's pick up where we left off after "Preparing Tabular Data for Training" on page 35, in which you completed the model training. Run these import statements again before you continue:

```
import functools
import numpy as np
import tensorflow as tf
import pandas as pd
from tensorflow import feature_column
from tensorflow.keras import layers
from sklearn.model_selection import train_test_split
import pprint
import tensorflow_model_analysis as tfma
from google.protobuf import text_format
```

Once the model is trained (which you did in Chapter 3), you can use it to predict the survival probability of each passenger in the test dataset:

```
prediction_raw = model.predict(test_ds)
```

This command, prediction_raw, produces a NumPy array with a probability value for each passenger:

```
array([[0.6699142 ],
       [0.6239286 ],
       [0.06013593],
.....
       [0.10424912]], dtype=float32)
```

Now let's make this array into a Python list and append it as a new column to the test dataset:

```
prediction_list = prediction_raw.squeeze().tolist()
test_df['predicted'] = prediction_list
```

The new column, called "predicted," is the last column. For visibility, you might want to reorder the columns by moving this last column to be first, next to the ground truth column, "survived."

```
# Put predicted as first col, next to survived.
cols = list(test_df.columns)
cols = [cols[-1]] + cols[:-1]
test_df = test_df[cols]
```

Now test_df looks like Figure 10-2, and we can easily compare the model's predictions to the real-life outcomes.

	predicted	survived	sex	age	n_siblings_spouses	parch	fare	class	deck	embark_town	alone
247	0.869914	1	male	32.0	0	0	56.4958	Third	unknown	Southampton	y
112	0.623929	0	female	20.0	0	0	8.6625	Third	unknown	Southampton	y
129	0.080136	0	male	28.0	0	0	0.0000	Second	unknown	Southampton	y
29	0.806499	1	female	28.0	1	1	22.3583	Third	F	Cherbourg	n
226	0.163873	0	male	18.0	0	0	7.7750	Third	unknown	Southampton	y
...	...	...	...	...	...	...	...	...	...	...	...
19	0.102746	0	male	29.0	0	0	8.0500	Third	unknown	Southampton	y
164	0.097697	1	male	32.0	0	0	7.9250	Third	unknown	Southampton	y
203	0.167355	0	male	42.0	0	0	7.6500	Third	F	Southampton	y
190	0.946212	0	male	58.0	0	2	113.2750	First	D	Cherbourg	n
171	0.104249	0	male	28.0	0	0	7.8958	Third	unknown	Southampton	y

106 rows × 11 columns

Figure 10-2. Titanic test dataset with predictions appended

Fairness Evaluation

To investigate the fairness of your model's predictions, you need a good understanding of your use case and the data you've used for training. Just looking at data alone probably won't give you enough context, background, or situational awareness to investigate fairness or model bias. Therefore, it is paramount that you understand the use case, the purpose of the model, who is using it, and the potential real-world impact should the model get the prediction wrong.

The *Titanic* had three cabin classes: first class was the most expensive, second class was in the middle, and third class, or steerage, was the least expensive and in the lower decks. It is well documented that most passengers who survived were female and in first-class cabins. We also know that gender and class played an important role in the selection process for getting on the lifeboats. That selection process prioritized women and children over men. Because this background is so well known, this dataset is suitable as a didactic example to investigate model bias.

When making predictions, as I mentioned in the introduction, a model inevitably recapitulates any bias or imbalance in the training data. So an interesting question to lead off our investigation would be this: how well does the model predict passenger survival in different genders and classes?

Let's start with the following code block for eval_config, which defines our investigation:

```
eval_config = text_format.Parse("""          ❶
  model_specs {                              ❷
    prediction_key: 'predicted',
    label_key: 'survived'
  }
  metrics_specs {                            ❸
    metrics {class_name: "AUC"}
    metrics {
      class_name: "FairnessIndicators"
      config: '{"thresholds": [0.1, 0.50, 0.90]}'
    }
    metrics { class_name: "ExampleCount" }
  }

  slicing_specs {                            ❹
    feature_keys: ['sex', 'class']
  }
  slicing_specs {}
  """, tfma.EvalConfig())                    ❺
```

❶ The eval_config object has to be formatted as a protobuf data type, which is why you needed to import text_format to parse the definition.

❷ Specify model_specs to document the two columns that you want to compare: "predicted" and "survived."

❸ Define the metrics for classification accuracy and the three thresholds for assigning survival status based on survival probability.

❹ This is where you declare which features you want to use to investigate model bias. feature_keys in slicing_specs holds a list of features to examine for bias: in our case, "sex" and "class." Since there are two unique values for sex and three different classes, the fairness indicators will evaluate model bias for six different interaction groups. If only one feature were listed, the fairness indicators would evaluate bias on that feature alone.

❺ All this information is wrapped inside """" """" triple double quotes, which makes it a plain-text representation. This text string is merged into a `tfma.EvalConfig` message.

Now define an output path to store the fairness indicator's results:

```
OUTPUT_PATH = './titanic-fairness'
```

Then run the model analysis routine:

```
eval_result = tfma.analyze_raw_data(
  data= test_df,
  eval_config=eval_config,
  output_path=OUTPUT_PATH)
```

From the preceding code, you can see that `test_df` is the test data, with prediction added to it. You will use the `tfma.analyze_raw_data` function to perform the fairness analysis.

TIP

If you are running this example in a local Jupyter Notebook, you need to enable the Jupyter Notebook to display the Fairness Indicators GUI. In the next cell, input the following command:

```
!jupyter nbextension enable tensorflow_model_analysis
--user -py
```

If you are using a Google Colab notebook, this step is not necessary.

Rendering Fairness Indicators

Now you are ready to take a look at your `eval_result`. Run this command:

```
tfma.addons.fairness.view.widget_view.render_fairness_indicator(
  eval_result)
```

You will see the Fairness Indicators interactive GUI in your notebook (see Figure 10-3).

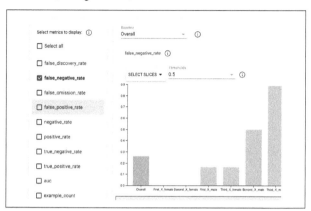

Figure 10-3. Fairness Indicators interactive GUI

In this figure, the metric of false-positive rate is selected.[1] In the context of the *Titanic* dataset, a *false positive* means that the model predicted the passenger would survive, but in reality they did not.

Let's take a look at this GUI. In the left panel (as shown in Figure 10-3), you have all the metrics available. In the right panel, you see a bar chart for the selected metric. Feature combinations (sex and class) in the `slice_specs` are indicated.

At a threshold of 0.5 (where a probability greater than 0.5 is considered a positive—that is, survival), the false-positive rate is high for male passengers in first class and female passengers in second and third class.

---

1 If you need a refresher on classification metrics, I suggest the succinct and useful review included in Google's *Machine Learning Crash Course* (*https://oreil.ly/yEEdm*).

In the table below the bar chart, you have a more detailed view of actual metric and sample size (see Figure 10-4).

feature	false_positive_rate@ 0.5	false_positive_rate@ 0.5 against Overall	example count
Overall	0.217	0%	106
class_X_sex:First_X_female	NaN	NaN%	13
class_X_sex:First_X_male	0.5	130.769%	16
class_X_sex:Second_X_female	1	361.538%	9
class_X_sex:Second_X_male	0	-100%	13
class_X_sex:Third_X_female	0.875	303.846%	14
class_X_sex:Third_X_male	0	-100%	41

Figure 10-4. The Fairness Indicators dashboard showing the false-positive summary

Unsurprisingly, the model correctly predicts that all first-class female passengers survived, as you can see from the ground truth ("survived") column.

So why do second-class and third-class male passengers produce a false-positive rate of 0? Let's look at the actual outcome from test_df to understand why. Execute the following command to select male passengers in second-class accommodations:

```
sel_df = test_df[(test_df['sex'] == 'male') & (test_df['class'] ==
'Second')]
```

Then display `sel_df`:

```
sel_df
```

Figure 10-5 shows the resulting list of all second-class male passengers. Let's use it to look for false positives. Is there any passenger in this group who did not survive (that is, the "survived" column displays 0) for whom the model predicted a probability greater than the threshold (0.5)? No. So Fairness Indicators states that the rate of false positives is 0.

	predicted	survived	sex	age	n_siblings_spouses	parch	fare	class	deck	embark_town	alone
129	0.060136	0	male	28.00	0	0	0.0000	Second	unknown	Southampton	y
67	0.218701	0	male	54.00	1	0	26.0000	Second	unknown	Southampton	n
244	0.867168	1	male	1.00	0	2	37.0042	Second	unknown	Cherbourg	n
82	0.270551	0	male	24.00	0	0	10.5000	Second	unknown	Southampton	y
246	0.673106	1	male	0.83	1	1	18.7500	Second	unknown	Southampton	n
4	0.335883	1	male	34.00	0	0	13.0000	Second	D	Southampton	y
64	0.200719	0	male	29.00	0	0	10.5000	Second	unknown	Southampton	y
37	0.182209	0	male	42.00	0	0	13.0000	Second	unknown	Southampton	y
254	0.268147	0	male	21.00	1	0	11.5000	Second	unknown	Southampton	n
77	0.182209	1	male	42.00	0	0	13.0000	Second	unknown	Southampton	y
211	0.200846	0	male	34.00	0	0	13.0000	Second	unknown	Southampton	y
223	0.332330	0	male	18.00	0	0	11.5000	Second	unknown	Southampton	y
189	0.322336	0	male	23.00	0	0	13.0000	Second	unknown	Southampton	y

Figure 10-5. The Titanic test dataset's second-class male passengers

You may notice that some second-class male passengers did survive, yet the model predicted their probability of survival as less than the threshold of 0.5. These are known as *false-negative* cases: they actually survived, but the model predicted they didn't. In short, they beat the odds! So let's uncheck the false-positive metric and check our false-negative metric to see what it looks like across gender and class combinations (see Figure 10-6).

As you can see in Figure 10-6, at the current threshold of 0.5, men and boys in second and third class have a very high false-negative rate.

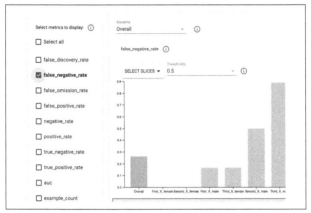

Figure 10-6. The Titanic test dataset's false-negative metric

This exercise makes it clear that the model exhibits bias with regard to gender and class. The model performs best for first-class female passengers, which reflects the obvious skew in the data: nearly all the first-class female passengers survived, while everybody else's chances are not so clear-cut. You can examine the training data to confirm this skew with the following statement, which groups training data by the "sex," "class," and "survived" columns:

```
TRAIN_DATA_URL =
https://storage.googleapis.com/tf-datasets/titanic/train.csv
train_file_path = tf.keras.utils.get_file("train.csv",
TRAIN_DATA_URL)
train_df = pd.read_csv(train_file_path,
header='infer')
train_df.groupby(['sex', 'class', 'survived' ]).size().
reset_index(name='counts')
```

This will produce the summary shown in Figure 10-7.

As you can see, of the 69 first-class female passengers, only two died. You trained your model based on this data, so it is very easy for the model to learn that as long as the class and sex indicators are first and female, it should predict a high probability of survival. You can also see that female passengers in

---

third class, where 52 survived and 41 died, did not enjoy the same favorable outcome. This is a classic case of an imbalanced data label (the "survived" column) leading to model bias. Of all the possible gender–class combinations here, no other combination has odds as good as those of female passengers in first class. This made it more challenging for the model to get the prediction right.

	sex	class	survived	counts
0	female	First	0	2
1	female	First	1	67
2	female	Second	0	5
3	female	Second	1	50
4	female	Third	0	41
5	female	Third	1	52
6	male	First	0	56
7	male	First	1	34
8	male	Second	0	64
9	male	Second	1	8
10	male	Third	0	216
11	male	Third	1	32

Figure 10-7. Summary of survival count by passenger group in the Titanic training dataset

When using Fairness Indicators, you can toggle the threshold drop-down box to display metrics for the three different thresholds in the same bar chart, as shown in Figure 10-8.

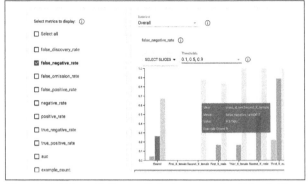

Figure 10-8. Fairness Indicators displaying metrics for all thresholds

For readability, each threshold is color coded. You can hover your cursor mouse over a bar to ascertain the color assignment. This helps you interpret the impact of different thresholds on each metric.

From this example, we can draw the following conclusions:

- Passenger features such as gender and cabin class (as a stand-in for socioeconomic class) primarily determine the outcome.

- The model is much more accurate for first- and second-class female passengers than for any other groups, which strongly suggests that the model bias is driven by gender and class.

- Survival probability favors a certain gender and certain classes. This is known as *data imbalance*, and it is consistent with historical accounts of what happened.

- Interactions between features such as gender and class (called *feature cross*) reveal a deeper level of model bias. Women and girls in the lowest (third) class did not have a favorable survival outcome—neither in the training data and model accuracy nor, tragically, in real life.

You may be tempted to try to tweak your model architecture or training strategies when you see model bias. But the underlying problem here is the imbalance in the training data, which reflects real-life unequal outcomes based on gender and class. Without bringing about a more equitable outcome for the other gender and classes, fixing the model architecture or training strategies is not going to create a fair model.

This example is a relatively simple one: the hypothesis here is that examining gender and class may reveal model bias and is therefore worth investigating. Because it is such a widely discussed historical event, most people have at least some awareness of the context and events leading to the *Titanic* tragedy. However, in many situations, formulating a hypothesis to investigate potential bias in data will not be straightforward. Your data might not contain personal attributes or personally identifiable information (PII), and you might not have the domain knowledge or contextual understanding of where and how the training data was collected and how this might affect the outcomes. Data scientists and ML model developers should definitely collaborate with subject matter experts to contextualize their training data. Sometimes, if model fairness is a primary concern, removing the features in question is the best and most sensible option.

Hyperparameter Tuning

Hyperparameters are values or attributes designated to control the model training process. The idea is that you want to train the model with combinations of these values and determine the best combinations. The only way to know which combination is best is to try these values out, so you want an efficient way to iterate all combinations and narrow down the choices to the best one. Hyperparameters are often applied to model architecture, such as the number of nodes in a dense layer of the deep-learning neural network. Hyperparameters can be applied to the training routine, such as an optimizer that performs back propagation (which you learned about in Chapter 8) during the training process. They can also be applied to the *learning rate*, which specifies how much you want to update the model with incremental changes in weights and bias, which in turn determines how big a step it is for back propagation to update weights and bias during training. Taken together, hyperparameters can be numeric (number of nodes, learning rate) or nonnumeric (name of optimizer).

As of TensorFlow distribution 2.4.1, the Keras Tuner library was not yet part of the standard distribution. That means you'll need to install it. You can run this installation statement in your command terminal or Google Colab notebook:

```
pip install keras-tuner
```

If running it within a Jupyter Notebook cell, use this version:

```
!pip install keras-tuner
```

The exclamation point (!) tells the Notebook cell to interpret this as a command rather than as Python code.

Once the installation is complete, you will import it as usual. Notice that while the library name is hyphenated when you install it, the hyphen is not used in the import statement:

```
import kerastuner as kt
```

From the Keras Tuner's perspective, hyperparameters have three data types: integer, item choice (a list of discrete values or objects), and floating point.

Integer Lists as Hyperparameters

It's easier to see how to use Keras Tuner through examples, so let's say you want to try different numbers of nodes in a dense layer. First you'll define the possible numbers, then pass that list to the dense layer's definition:

```
hp_node_count = hp.Int('units', min_value = 16, max_value = 32,
step = 8)
tf.keras.layers.Dense(units = hp_node_count, activation = 'relu')
```

In the preceding code, `hp.Int` defines an alias: `hp_node_count`. This alias contains a list of integers (16, 24, and 32), which you pass to the `Dense` layer as `units`. Your goal is to see which number works the best.

Item Choice as Hyperparameters

Another way to set hyperparameters is to put all selections in a list as discrete items, or choices. This is accomplished with the `hp.Choice` function:

```
hp_units = hp.Choice('units', values = [16, 18, 21])
```

Here is an example of specifying an activation function by names:

```
hp_activation = hp.Choice('dense_activation',
                values=['relu', 'tanh', 'sigmoid'])
```

Floating-Point Values as Hyperparameters

In many cases, you'll want to try different decimal values in the training routine. This is very common if you want to select a learning rate for the optimizer. To do so, use this command:

```
hp_learning_rate = hp.Float('learning_rate',
 min_value = 1e-4,
 max_value = 1e-2,
 step = 1e-3)
optimizer=tf.keras.optimizers.SGD(
 lr=hp_learning_rate,
 momentum=0.5)
```

End-to-End Hyperparameter Tuning

Hyperparameter tuning is a time-consuming process. It involves trying multiple combinations, with each combination trained to the same number of epochs. For a "brute force" approach, you would have to loop (iterate) through each combination of hyperparameters and *then* launch the training routine.

The advantage of using Keras Tuner is its early-stop implementation: if a particular combination doesn't seem to improve the results, it will terminate the training routine and move to the next combination. This helps reduce total training time.

Next, you're going to see how to perform and optimize hyperparameter searches using a strategy known as *hyperband search*. Hyperband search utilizes the principle of successive reduction during training. In each loop through all hyperparameter combinations, the algorithm ranks how well the model performed for all combinations and discards the worse half of the parameter combinations. The better half of combinations will receive more processor cores and memory in the next round. This continues until the last combination remains, eliminating all but the best hyperparameter combination.

It's a bit like a playoff bracket (*https://oreil.ly/jq8Lb*) in sports: each round and each matchup eliminates the lower-seed team. In hyperband search, though, the losing team is declared the loser *before the game completes*. This process continues until the championship matchup, where the number-one seed team is the eventual champion. This strategy is a lot less wasteful than the brute force approach, where each combination is trained to its full epochs, since that eats up a lot of training resources and time.

Let's apply what you've learned to the CIFAR-10 image classification dataset you worked with in the previous chapter.

Import Libraries and Load Data

I recommend using a Google Colab notebook with a GPU to run the code in this example. As of TensorFlow 2.4.1, the Keras Tuner is not yet a part of the TensorFlow distribution, so in the Google Colab environment, you need to run a pip install command:

```
pip install -q -U keras-tuner
```

Once it is installed, import it along with all the other libraries:

```
import tensorflow as tf
import kerastuner as kt
from tensorflow.keras import datasets, layers, models
import numpy as np
import matplotlib.pylab as plt
import os
from datetime import datetime
print(tf.__version__)
```

It will show the current version of TensorFlow—in this case, 2.4.1. Then load and normalize images in one cell:

```
(train_images, train_labels), (test_images, test_labels) =
datasets.cifar10.load_data()

# Normalize pixel values to be between 0 and 1
train_images, test_images = train_images / 255.0,
test_images / 255.0
```

Now provide a list of labels:

```
# Plain-text name in alphabetical order.
https://www.cs.toronto.edu/~kriz/cifar.html
CLASS_NAMES = ['airplane', 'automobile', 'bird', 'cat',
              'deer','dog', 'frog', 'horse', 'ship', 'truck']
```

Next, convert the images to datasets by merging images and labels into tensors. Split the test dataset into two groups—the first 500 images (for validation during training), and everything else (for testing):

```
validation_dataset = tf.data.Dataset.from_tensor_slices(
 (test_images[:500], test_labels[:500]))

test_dataset = tf.data.Dataset.from_tensor_slices(
 (test_images[500:], test_labels[500:]))

train_dataset = tf.data.Dataset.from_tensor_slices(
 (train_images, train_labels))
```

To ascertain the sample size of each dataset, execute the following commands:

```
train_dataset_size = len(list(train_dataset.as_numpy_iterator()))
print('Training data sample size: ', train_dataset_size)

validation_dataset_size = len(list(validation_dataset.
as_numpy_iterator()))
print('Validation data sample size: ', validation_dataset_size)
```

```
test_dataset_size = len(list(test_dataset.as_numpy_iterator()))
print('Test data sample size: ', test_dataset_size)
```

You should get something like the following output:

```
Training data sample size:  50000
Validation data sample size:  500
Test data sample size:  9500
```

Next, in order to take advantage of distributed training, define a `MirroredStrategy` object to handle distributed training:

```
strategy = tf.distribute.MirroredStrategy()
print('Number of devices: {}'.format(
strategy.num_replicas_in_sync))
```

In your Colab notebook, you should see the following output:

```
Number of devices: 1
```

Now set up sample batch parameters:

```
BUFFER_SIZE = 10000
BATCH_SIZE_PER_REPLICA = 64
BATCH_SIZE = BATCH_SIZE_PER_REPLICA * strategy.num_replicas_in_sync
STEPS_PER_EPOCH = train_dataset_size // BATCH_SIZE_PER_REPLICA
VALIDATION_STEPS = 1
```

Shuffle and batch all datasets:

```
train_dataset = train_dataset.repeat().shuffle(BUFFER_SIZE).batch(
BATCH_SIZE)
validation_dataset = validation_dataset.shuffle(BUFFER_SIZE).batch(
validation_dataset_size)
test_dataset = test_dataset.batch(test_dataset_size)
```

Now you can create a function to wrap the model architecture:

```
def build_model(hp):
  model = tf.keras.Sequential()
  # Node count for next layer as hyperparameter
  hp_node_count = hp.Int('units', min_value=16, max_value=32,
      step=8)
  model.add(tf.keras.layers.Conv2D(filters = hp_node_count,
      kernel_size=(3, 3),
      activation='relu',
      name = 'conv_1',
      kernel_initializer='glorot_uniform',
      padding='same', input_shape = (32,32,3)))
  model.add(tf.keras.layers.MaxPooling2D(pool_size=(2, 2)))
```

```
model.add(tf.keras.layers.Flatten(name = 'flat_1'))
# Activation function for next layer as hyperparameter
hp_AF = hp.Choice('dense_activation',
    values = ['relu', 'tanh'])
model.add(tf.keras.layers.Dense(256, activation=hp_AF,
    kernel_initializer='glorot_uniform',
    name = 'dense_1'))
model.add(tf.keras.layers.Dense(10,
    activation='softmax',
    name = 'custom_class'))

model.build([None, 32, 32, 3])
# Compile model with optimizer
# Learning rate as hyperparameter
hp_LR = hp.Float('learning_rate', 1e-2, 1e-4)

model.compile(
    loss=tf.keras.losses.SparseCategoricalCrossentropy(
        from_logits=True),
    optimizer=tf.keras.optimizers.Adam(
        learning_rate=hp_LR),
    metrics=['accuracy'])

return model
```

There are some major differences between this function and the one you saw in Chapter 9. The function now expects an input object, hp. This means the function will be invoked by a hyperparameter tuning object named hp.

In the model architecture, node count for the first layer conv_1 is declared for hyperparameter search by using hp_node_count. An activation function for layer dense_1 is also declared for hyperparameter search by using hp_AF. Finally, the learning rate in optimizer is declared for hyperparameter search by using hp_LR. This function returns the model with hyperparameters declared.

Next, define an object (tuner) with kt.Hyperband, which takes the build_model function as an input:

```
tuner = kt.Hyperband(hypermodel = build_model,
                     objective='val_accuracy',
                     max_epochs=10,
                     factor=3,
                     directory='hp_dir',
                     project_name='hp_kt')
```

You pass the following inputs to define the `tuner` object:

`hypermodel`
> A function that defines the model architecture and optimizer.

`objective`
> A training metric used to evaluate model performance.

`max_epochs`
> The maximum number of epochs for model training.

`factor`
> The reduction factor for epochs and the number of models in each bracket. Models ranked in the top 1/factor are selected and advanced to the next round of training. If the factor is 2, then the top half will advance to the next round. If the factor is 4, then the top quarter will advance to the next round.

`directory`
> The target directory to write results to, such as checkpoints for each model.

`project_name`
> The prefix for all files saved in the target directory.

Here you can define an early stop to stop training if there is no improvement in validation accuracy for five epochs:

```
early_stop = tf.keras.callbacks.EarlyStopping(
 monitor='val_accuracy',
 patience=5)
```

Now you are ready to launch the search via the Hyperband algorithm. It will print out the best hyperparameters when the search is complete:

```
tuner.search(train_dataset,
             steps_per_epoch = STEPS_PER_EPOCH,
             validation_data = validation_dataset,
             validation_steps = VALIDATION_STEPS,
             epochs = 15,
             callbacks = [early_stop]
             )
```

```
# Get the optimal hyperparameters
best_hps=tuner.get_best_hyperparameters(num_trials=1)[0]

print(f"""
The hyperparameter search is complete. The optimal number of units
in conv_1 layer is {best_hps.get('units')} and the optimal
learning rate for the optimizer is {best_hps.get('learning_rate')}
and the optimal activation for dense_1 layer
is {best_hps.get('dense_activation')}.
""")
```

As you can see, after the search, best_hps holds all the information about the best hyperparameter values.

When you run this example in a Colab notebook with a GPU, it usually takes about 10 minutes to complete. Expect to see output that looks something like this:

```
Trial 42 Complete [00h 01m 14s]
val_accuracy: 0.593999981880188

Best val_accuracy So Far: 0.6579999923706055
Total elapsed time: 00h 28m 53s
INFO:tensorflow:Oracle triggered exit

The hyperparameter search is complete. The optimal number of units
in conv_1 layer is 24 and the optimal learning rate for the
optimizer is 0.0013005004751682134 and the optimal activation
for dense_1 layer is relu.
```

This output tells us that the best hyperparameters are as follows:

- Optimal node count for layer conv_1 is 24.

- Optimal learning rate for optimizer is 0.0013005004751682134.

- Optimal activation function choice for dense_1 is "relu."

Now that you have the best hyperparameters, you need to formally train the model with these values. The Keras Tuner has a high-level function called hypermodel.build that makes this a single-command process:

```
best_hp_model = tuner.hypermodel.build(best_hps)
```

After that, set up the checkpoint directory as you did in Chapter 9:

```
MODEL_NAME = 'myCIFAR10-{}'.format(datetime.datetime.now().
strftime("%Y%m%d-%H%M%S"))
print(MODEL_NAME)
checkpoint_dir = './' + MODEL_NAME
checkpoint_prefix = os.path.join(
checkpoint_dir, "ckpt-{epoch}")
print(checkpoint_prefix)
```

You'll also set up the checkpoint the same way you did in Chapter 9:

```
myCheckPoint = tf.keras.callbacks.ModelCheckpoint(
    filepath=checkpoint_prefix,
    monitor='val_accuracy',
    mode='max',
    save_weights_only = True,
    save_best_only = True
    )
```

Now it's time to launch the model training process with the best_hp_model object:

```
best_hp_model.fit(train_dataset,
            steps_per_epoch = STEPS_PER_EPOCH,
            validation_data = validation_dataset,
            validation_steps = VALIDATION_STEPS,
            epochs = 15,
            callbacks = [early_stop, myCheckPoint])
```

Once the training is complete, load the model that has the highest validation accuracy. With save_best_only set to True, the best model will be the one in the latest checkpoint:

```
best_hp_model.load_weights(tf.train.latest_checkpoint(
checkpoint_dir))
```

Now best_hp_model is ready for serving. It is trained with the best hyperparameters and the weights and bias are loaded from the best training epoch, which is the one that yielded the highest validation accuracy.

Wrapping Up

In this chapter, you learned how to improve your model building and quality assurance processes.

One of the most frequent and important quality assurance concerns for ML models is fairness. Fairness Indicators is a tool that can help you investigate model bias across many different feature interactions and combinations. When evaluating model fairness, you have to look for model bias in the training data. You also need to rely on subject matter experts for context as you develop your hypothesis to investigate any model bias. In the *Titanic* example this process is pretty straightforward, because it is obvious that gender and class played important roles in determining each individual's chance for survival. However, in practice, there are many other factors that complicate matters, including how the data was collected and whether or not the context or conditions for data collection favored one group within the sample source over others.

In the model building process, hyperparameter tuning is time consuming. In the past, you had to iterate over each combination of potential hyperparameter values to search for the best combination. With the Keras Tuner library, however, a relatively advanced search algorithm known as Hyperband conducts searches efficiently using a tournament-bracket style framework. In this framework, models trained on weak hyperparameters are terminated and removed before the training epochs complete. This reduces total search time and the best hyperparameters emerge as the winner. All you need to do is formally train the same model with the winning combination.

With this knowledge, you are now ready to elevate your TensorFlow model development and testing skills to the next level.

Index

About the Author

KC Tung is a cloud solution architect at Microsoft who specializes in designing and delivering machine learning and AI solutions in an enterprise cloud architecture. He helps enterprise customers with use-case driven architecture, AI/ML model development/deployment in the cloud, and technology selection and integration best suited for their requirements. He is a Microsoft-certified AI engineer and data engineer. He holds a PhD in molecular biophysics from the University of Texas Southwestern Medical, and has spoken at the 2018 O'Reilly AI Conference in San Francisco and the 2019 O'Reilly Tensorflow World Conference in San Jose.

Colophon

The animal on the cover of *TensorFlow 2 Pocket Reference* is a blue tilapia (*Oreochromis aureus*), native to fresh and brackish water systems of northern and western Africa and the Middle East.

Blue tilapia are a blue-tinged fish with a paler belly. Adults average over a foot long and weigh up to five or six pounds. For the most part they consume plants but will also eat plankton, and the young will consume invertebrates. They can live up to ten years in captivity.

These fish are mouthbrooders: after laying her eggs (numbering a few dozen to a hundred at once), a female then protectively takes the eggs into her mouth until they hatch a few days later. The newly hatched fish, called fry, will feed from secretions on the insides of their mother's mouth. These secretions can pass along immunity to disease to a fry from its mother.

Because tilapia grow quickly, eat inexpensive plant material, and are amenable to crowded living conditions, they're well-suited for fish farming, and they are now farmed in over a hundred countries. With the health benefits of eating fish driving demand for fish worldwide, tilapia consumption has grown exponentially: in recent years, global consumption of this fish amounted to over six million tons. Tilapia are now among the most commonly consumed fish in the world.

Blue tilapia has been introduced as a food fish into waterways around the world, including in many parts of the United States, where it has been attested as a destructive invasive species, causing declines in native fish and mussel species. Many of the animals on O'Reilly covers are endangered; all of them are important to the world.

The color cover illustration is by Karen Montgomery, based on a black and white engraving from Richard Lydekker's *Royal Natural History* (1893). The cover fonts are Gilroy Semibold and Guardian Sans. The text font is Adobe Minion Pro; the heading font is Adobe Myriad Condensed; and the code font is Dalton Maag's Ubuntu Mono.

Milton Keynes UK
Ingram Content Group UK Ltd.
UKHW020159270724
446142UK00009B/151